T0219394

How We Teach Science

How We Teach Science

What's Changed, and Why It Matters

JOHN L. RUDOLPH

Harvard University Press

Cambridge, Massachusetts
London, England
2019

Second printing

Library of Congress Cataloging-in-Publication Data

Names: Rudolph, John L., 1964– author.
Title: How we teach science : what's changed, and why it matters /
    John L. Rudolph.
Description: Cambridge, Massachusetts : Harvard University Press,
    2019. | Includes bibliographical references and index.
Identifiers: LCCN 2018037906 | ISBN 9780674919341 (alk. paper)
Subjects: LCSH: Science—Study and teaching (Secondary)—
    United States—History. | Science—Methodology—Study and
    teaching (Secondary)—United States—History. | Education—
    Social aspects—United States—History.
Classification: LCC Q183.3.A1 R828 2019 | DDC 507.1/273—
    dc23 LC record available at https://lccn.loc.gov/2018037906

For Connie and Walter

CONTENTS

How We Teach Science

Introduction

FEW SUBJECTS ENJOY as much status in the American high school as the sciences. When international crises strike, economic downturns overtake the nation, or high-tech companies are squeezed by global competition, a common response from legislators, policymakers, and other leaders is to double down on science education. They call for more rigorous standards, new curricula, increased funding for laboratory equipment, and better preparation of teachers in the key disciplines of biology, chemistry, and physics. High school subjects such as these are often viewed as gateways to college and, for some, a path to lucrative careers and a comfortable middle-class life. There's no problem that seemingly can't be solved by channeling more and more students into the science education pipeline. What happens in that pipeline, however, particularly at its widest point in the high schools of the United States—where nearly all students are given their earliest exposure to science—has rarely been considered in much detail.

This book aims to provide that detailed look, particularly at how what's happened in science classrooms has changed over the years and what the consequences of those changes might be for how we relate to science as citizens. No one would doubt, certainly, that students are taught something different now compared to what they were taught nearly two centuries ago. But the most significant changes have come not in the way most of us might think. To be sure, the steady march of scientific research over the years has added ever more content knowledge to student textbooks. The latest concepts, models, and theories about the natural world are now taught in place of the less refined ideas of the past—and so there's more to know and more accurate science for students to master than there used to be. Along with all the newer content, however, is what's been taught about where that content has come from—in other words, what can be called the process of science. This, surprisingly, has assumed dramatically different forms in science classrooms over the years.

Let's start with the scientific method, something we're all familiar with. Many of us can recite (or at least are aware of) the version that includes the handful of steps that purportedly guide scientists in their work—define a problem, form a hypothesis, gather data, draw a conclusion, et cetera. It's part of our shared understanding. There was a time when it was common for scientists—or nearly anyone, in fact—to make regular, sometimes emphatic pleas for more people to use this seemingly infallible method in their personal lives as well as for addressing broader civic problems. Interestingly enough, however, such methodological appeals are no longer a part of our public discourse. Yet the idea of a scientific method remains a fixture of our common cultural heritage nonetheless. Whenever the question arises of how science achieves its many, often remarkable accomplishments, the explanation to which nearly all of us default is: "Why, one follows the scientific method, of course."[1]

As a career science educator—first as an eighth-grade teacher years ago and now as a professor of science education—the source of the step-by-step version of how science works is obvious. It has long been a staple of what teachers teach when they teach about science. Indeed, I have spent the better part of a career trying to correct this simplistic account of the process of science. I myself was taught the steps of the scientific method as a student in southeastern Wisconsin in the late 1970s and early 1980s, and over the years those steps came to be as familiar to me then as they continue to be for many students today. So when I began studying to be a science teacher in college, I was surprised to learn that there was no such thing as the scientific method. That was certainly the message my science-teaching methods professor instilled in us. He taught that the focus of our teaching instead should be on something called "scientific inquiry," the details and complexities of which we learned about from an early 1960s essay by a University of Chicago professor by the name of Joseph Schwab.

I dutifully accepted this novel (at least to me) characterization of the scientific process and set out for various jobs across Wisconsin teaching middle school science, then high school physics and chemistry, and finally biology. My commitment to science education led me eventually to graduate school and then to a faculty position running a teacher certification program where, in my own science-teaching methods course, I too deliberately sought to eradicate the unthinking acceptance of *the* scientific method, to correct the widespread public misconception about how the process of science unfolds in actual practice. But even with explicit and repeated instruction on this very point, the pre-service teachers in my class held tight to the scientific method they had internalized from school. I vividly remember visiting a local high school one day to observe a student teacher working in the classroom of one of our program's recent graduates. As I entered her room, I saw displayed on the front wall a poster listing in large letters the canonical steps of the scientific method. This was as damning an assessment of my teaching as any I might have encountered.

What struck me more than anything during that visit, in retrospect, was that there clearly was something important about schools as sites for teaching the methods of science. Without question, teachers valued an understanding of the scientific process as an instructional goal. Yet it was also clear that there was something about science classrooms or science teaching more generally that was responsible for reproducing this simplistic, stepwise routine for how scientists conducted research.

What went on in that high school science classroom was seemingly inconsequential—few of us care all that much about what gets taught in science classes (the topics of evolution and human reproduction being notable exceptions). Yet at the same time, as I came to realize, the teaching of the scientific method in that space held tremendous significance. Here, after all, was the one place where the public is taught what science is and how it operates to produce what we all accept, more or less, as the truths of the natural world. Almost from the beginning of mass schooling in the nineteenth century, when the subjects of physics, chemistry, botany, and zoology became a standard part of the typical secondary school pupil's studies, the methods employed in those disciplinary fields have been a central part of their teaching. Understanding the nature of the scientific method (or the methods of science), in one form or another, has been one of the most enduring goals of science education, one of the key learning outcomes that science educators, business leaders, and policymakers have repeatedly and enthusiastically endorsed. In our current era, the National Academy of Sciences has continued to call for science teaching to include a strong emphasis on *how* scientists come to know, rather than just *what* is known. Not only is science a body of content knowledge, the National Academy asserts, but "it is also a set of practices used to establish, extend, and refine that knowledge." In this view, understanding science requires knowing about the *process* of knowledge construction.[2]

But while attention to the process of science (to use perhaps the broadest descriptive phrase) has been an established element of science classrooms over the past 130 years, the characterization of that process and how it has been taught to students has been far from static. Accounts of how science generates knowledge have turned over repeatedly since they were first introduced into the school curriculum in the 1880s. One of the more significant challenges in writing this book, in fact, has been trying to distinguish and keep track of the various ways that the process of science has been described. With respect to "the scientific method," for example, there are some readers (typically those with some firsthand knowledge of scientific work) who will exclaim—as my college professor did so many years ago—that no such thing exists! Other readers (even some practicing scientists) will see no problem using the phrase as a shorthand for scientific thinking. Much depends on the particular referent each reader has in mind.

In schools, explicitly teaching the scientific method as a universal process was common in the early and middle decades of the twentieth century. Before that, however, a less well-defined scientific method showed up in schools as the "laboratory method" of instruction (though still at times called the scientific method or the inductive method), which was implemented hand in hand with apparatus made available in newly constructed school laboratories of the day. In another turn among many, the latter decades of the twentieth century saw the stepwise scientific method replaced by teaching science as the process of "inquiry" that I encountered in college. And more recently still one finds educators advocating the "practices" of science in place of scientific inquiry and method. All, though, are attempts to make sense of and describe the manner in which science arrives at what most of us accept as reliable knowledge about the world, the facts of science.

Tracing these shifting curricular and popular accounts (of which there are numerous variants and hybrids) is precisely the point of this work, and it illustrates the remarkable plasticity of the image of scientific work as packaged for classroom consumption. Indeed, over the 130 years that constitute the duration of this story, classroom portrayals of the scientific process have varied far more than the practice of science itself, which has remained remarkably stable, for the most part, over the same period.

Why Method Matters

The question of how different educators and reformers have sought to portray the methods of science is the particular subject of this book. But why should we care that these portrayals changed from one period to the next or that there might be multiple ways to present the process of science even today? As it turns out, it matters a great deal. The way students encounter science in classrooms has consequences not only for science itself but also for society as a whole. Students, after all, make up the future citizenry of the United States. What students experience in schools inevitably shapes perceptions of how science is done and, in turn, influences the relationship between science and the public. Let me explain.

Members of the general public rarely encounter descriptions of how science works once they leave the classroom and busy themselves with their everyday affairs: going to work, raising families, engaging in recreational pursuits. Such information is typically missing even from scientific journals and research papers. Among scientists, methodology is simply a non-topic. The "methods" section in a scientific paper typically covers only the particular techniques or protocols employed in a study; researchers rarely share the general inductive, deductive, or abductive steps that they took to arrive at the paper's conclusions. The logical, evidentiary, or epistemological "rules" (if they can even be called that) scientists follow in any given inquiry are mostly tacit. This isn't to say that the topic of how scientists produce knowledge never comes up for debate; it does, certainly among philosophers of science and others who study science. But in everyday public discourse, descriptions of the scientific process appear only episodically, usually when something about scientific knowledge or science itself is at issue.[3]

Discussions of scientific methodology typically erupt publicly when the authority or the legitimate scope of science is in conflict with other social or cultural norms, knowledge systems, or local claims. The sociologist Thomas Gieryn has referred to these moments as boundary disputes. In debates over what does and does not count as science, what gets ruled in (as science) is allowed the authority to decide what counts as true. While a boundary dispute centers on the question of where the line between science and non-science is drawn, the decision about where to make that demarcation almost always hinges on an interpretation of process or methodology. In early nineteenth-century Britain, for example, the fledgling scientific profession produced reams of writing on scientific method as it competed with the church over epistemological (and thus cultural) authority. More than a century later in the United States, scientists drew on arguments over the scope of scientific methods to debate whether or not the social sciences should be grouped under the same funding umbrella as the natural sciences. In each case, questions about what counted required those involved to dig into its underlying methods and epistemology to make a decision of consequence, be it whether to include science as an institution of cultural significance or to appropriate taxpayer dollars to certain types of research.[4]

In instances such as these, public officials or citizens participating in those decisions routinely make some effort to tutor themselves in the nuances of scientific process, whether through their own reading, lecture attendance, or hearing of expert testimony on how science operates, in order to inform their judgment about the matter at hand. Citizens on the outside following newspaper accounts, online sources, or television reports on science topics similarly, if less rigorously, come to their own conclusions about how such matters should be decided.[5]

No one, however, whether public official or ordinary citizen, comes to such decisions as a blank slate. People at every level bring with them their own distinct images of what science is and how it works. Some of these ideas about science come through mass media—from sources such as movies, television shows, and the internet.[6] More significant, though, is what students learn about the process of science in school, whether through explicit didactic instruction or through engagement with carefully designed laboratory exercises or other hands-on activities. Unlike the hit-and-miss impressions from mass media, these experiences are officially sanctioned, systematic, regularly occurring, and tested. As a result, images of scientific work encountered in schools have an intellectual inertia, a permanence that leaves a lasting impression in the public mind, and these images set the stage for nearly all the citizen-science encounters to come in life.

These baseline views of science naturally condition—individual by individual—public reaction to the ongoing operations of science in a variety of ways. Individuals who perceive science as fundamentally dependent on experimental manipulation, for instance, would likely doubt the credibility of claims made by advocates of evolutionary biology, a science that typically relies on knowledge derived from non-experimental methods in its exploration of biological change. Similar concerns might be voiced (and indeed have been) about the claims of climate scientists, who use historical trend data and computer modeling to help make sense of highly complex natural systems, changes in which have occurred over time horizons that exceed the limits of direct human observation. No simple experiments can be performed in these cases to determine the truth or falsity of their conclusions. To take another example, seeing scientific methods as having been perfected in the physical sciences might suggest to some that science as a whole should not stray from its home base in well-established research fields such as physics and chemistry.[7] Alternatively, the belief that the essence of science lies in methods of knowledge production that are freestanding and universal would support their use in nearly any venue where there is a significant problem to be solved, whether that problem falls within "hard science" domains or in more distant areas of public concern that involve issues of culture, social relations, politics, or even morality.

Not unrelated to this is the question of how views of scientific methods situate the public with respect to the scientific enterprise in democratic political systems. Since the professionalization of science in the early 1800s, the scientific community has occupied a unique position in the intellectual and social strata of Western culture. The technical expertise of scientists has earned them a measure of public respect and intellectual authority accorded few other professional groups, authority that has grown as society has become increasingly dependent on the instrumental capabilities of science for reasons of both national security and economic prosperity.[8] However, as will be evident in the chapters that follow, since World War II members of the scientific community have increasingly argued that the means by which they have achieved this technological mastery derives from indefinable elements of their expert practice rather than from more routine, specifiable methods. Defining science in this way—as something that can be known only from the *inside*—puts the lay public in a position of deference to scientific expertise and judgment and, thus, *outside* the sphere of influence over funding priorities and program direction. Consequently, public perceptions of the intricacies of scientific process are central to determining where the legitimate exercise of expert judgment leaves off and the rightful expression of public preferences regarding science research policy begins. In other words, here the question concerns how much the public needs to know about how science works in order to have a voice in determining the kind of that science gets done.

Why History Matters

Understanding how portrayals of scientific method have appeared in schools is central to the larger goal of learning how they function in mediating the relationship between science and the public. Any brief survey of the history of science education in the United States shows that these portrayals have been fluid, often changing dramatically from one historical period to the next.[9] Clearly there is a complex range of factors at work beyond simple attempts to mirror or authentically represent scientific practice itself, and these factors have pushed and pulled depictions of scientific process in schools from one form to another. Tracing the changes in these depictions—looking at how they have been developed, promoted, and implemented in historical context—offers insight into the various factors that influenced their construction and, in addition, helps us understand how the individuals involved have viewed the essential nature of scientific work and what the proper relationship between science and the public should be.

Throughout the history of American science education, three main themes have factored into scientists' and educators' efforts to capture the process of science in a form that might effectively be taught to high school students. These themes tell us not only about the changing portrayals of process in the classroom but also about changing views of science over time, broader ideas about the nature of teaching and learning (held by scientists and educators alike), and the appropriate function of educational institutions in American society. I'll take each in turn.

The equation of *scientific process* and *pedagogy* throughout the history of science education is the most powerful of the three themes. From the beginning of laboratory teaching in the late 1800s, scientist educators believed that the experience of student learning should be modeled on the process of knowledge generation in science. Once it became accepted that science as a way of knowing enjoyed privileged access to the truth about the natural world, it logically followed that they should incorporate key elements of the scientific process into their pedagogy. Learning about nature the way scientists did, through laboratory study, ensured that the individual student's knowledge would be certain (since it came directly from sense experience). Decades later, following the flurry of reforms after the Soviet launch of Sputnik, a similar argument was made—that engaging in the process of scientific thinking (scientific inquiry in this case) was the ideal pedagogical method. Such engagement was the only way one could develop a true appreciation for the process of science as undertaken by scientists. In both of these cases, as well as others since the beginning of formal science education, attempts to model science pedagogy on the elements of authentic scientific practice—to have students do science to learn science—have exerted a strong hold on educational thought.

The second theme in this story about scientific process relates to the *nature of schooling* and arises from the challenges reformers faced when attempting to implement their particular educational visions. Anytime reforms moved from the planning stage into the classroom, problems inevitably arose. The reality of two dozen or more students (of varying abilities and interests) shuffling every fifty minutes into and out of a science classroom that may or may not be properly equipped has always been a challenge. That challenge is complicated by the fact that the teachers leading instruction often themselves lacked a firm grasp of the content to be taught or are not always sympathetic to the vision of science being advanced. In every era, these factors—the students, the material conditions of the classroom, and the teachers—impinge on pedagogical ideals in ways that reformers never seemed to fully account for. Although this has been true for educational reforms throughout history, the case of science education, it seems, has experienced this trouble more acutely than other school subjects. This can be attributed, I would argue, to the particular material requirements of scientific work and the expertise differential that exists between the scientist and the classroom science teacher, especially with respect to knowing what counts as the process of science—something teachers almost always come to know only secondhand.[10]

The third and perhaps most important theme concerns the *social purpose of science education*. That purpose, as it relates to scientific method, has shifted depending on both larger historical factors as well as self-interested ones (that is, whose vision is being advanced). Pedagogical ideas about science coming from practicing scientists, who predominated during the era of the laboratory method in the late 1800s and the postwar period (the mid-1950s through the mid-1970s), were often drawn from disciplinary norms and structures. In the earlier period, the process of science, they believed, had the power to instill morality and discipline in the individual student, while in the latter period it was for the benefit of science as an institution. With science educators (those who more closely identified professionally with education rather than science, as was the case from the 1910s through the 1940s), instruction in the methods of science was closely connected to the personal needs and interests of the student. The idea then was that understanding the process of science would improve the experience of Americans by giving them the ability to reason more effectively, which would allow them to solve the everyday problems they encountered. In each case, the portrayal of method was shaped by the intended purpose of instruction.

These themes highlight the complexity of how ideas about science and learning are formed and implemented in practice. What will be apparent in the chapters that follow is that despite the best efforts of scientists, educational leaders, classroom teachers, and reformers of any kind, intended messages get transformed along the way by the factors I have outlined above (and more). Moreover, the transformed messages or ideas, in this case about how science generates knowledge, are taken up and repurposed over time, flowing from the scientists and education reformers to the classroom and back again.[11]

The Outline of the Story

The history of science teaching in the American high school can be thought of as spanning four distinct periods. Chapters 1 and 2 tell the story of the first period. I begin with the teaching of science in schools as an "information" subject, as it was often called at the time. During these years, from the middle part of the nineteenth century until the late 1880s, science gained increasing status in schools for what it provided in the way of utilitarian knowledge. The value it appeared to hold—knowledge about how things worked—translated directly into teaching through textbooks and learning by rote. However, American scientists, following their immersion in European research models and teaching laboratories, sought to transform college and high school science education in the United States to reflect what they believed to be a superior pedagogical method centered on process that would convey benefits of liberal education such as mental discipline and even virtue. The key figures in this period of transition were Harvard president Charles

Eliot and Harvard physicist Edwin Hall in the physical sciences and Ne-braska botanist Charles Bessey in the biological sciences. Together they worked to establish the laboratory method as the indisputable standard of science education through the early 1900s, and it was a standard the public thoroughly embraced, as evidenced by the marvelous laboratories built in the thousands of new schools that popped up across the country.

The laboratory method, however, came under fire in the early twentieth century from science educators who increasingly identified with the educa-tion profession that grew up with the expansion of public education during this time. This marked the transition to the second period of teaching about method, which I cover in Chapters 3, 4, and 5. These individuals, whom I call the progressive science educators, included people such as Charles Riborg Mann and John Woodhull in physics and, later, Otis Caldwell, Henry Lin-ville, and George W. Hunter in biology. They drew upon new ideas from the field of psychology advanced by G. Stanley Hall of Clark University in the 1890s to question the appropriateness of rigid laboratory instruction for the masses of students filling American high schools. Hall's work on the needs and interests of adolescents at the time prompted a reexamination of how well the laboratory method engaged students, which led to the development of a new approach that framed the scientific process in terms of individual problem solving rather than as an exercise in formal logic.

The work of the pedagogue and philosopher John Dewey, who assumed leadership of the progressive education movement around 1910, was the key to establishing an entirely new characterization of the scientific method. Long interested in seeing science used as a foundation for individual thinking and public deliberation, Dewey pushed for science education to focus on teaching about the methods of knowledge production rather than facts. His book *How We Think,* a primer of sorts for teachers that laid out the essential nature of reflective thought, became a bestseller following its publication in 1910 and led to the widespread adoption of the now well-known five-step scientific method. Seizing upon this characterization of method, progressive science educators advanced new courses, new methods of pedagogy, and en-tirely new ways to think about science education in the high school focused on problem solving in everyday life.

The third major period in the history of teaching scientific method in the high school, which I cover in Chapters 6, 7, and 8, arrived following World War II when scientists found themselves at the helm of a large and growing scientific research enterprise funded by a federal government increasingly reliant on technical expertise to ensure the security of the nation. In this period of immense public investment—which accelerated after the Soviet launch of Sputnik—members of the scientific elite believed that perceptions of science mattered tremendously to the continued support of their enterprise. The scientific method of the progressive science educators, especially the five-step method attributed to Dewey, was seen as having the potential to undermine public support for science. As a result, a handful of research scientists bankrolled largely by the National Science Foundation embarked on a massive program to remake science teaching methods and materials to project an image of scientific process—what many called "science as inquiry"—that, they felt, better reflected the real work of scientists (and that would likely ensure continued public funding of scientific research).

The last two chapters describe the fourth and final period of the story. These trace the fate of the inquiry approach through the lean years of the 1970s, when public esteem for science waned. Fears of a new economic challenge from Japan, however, prompted a fresh call for education reform aimed at preparing citizens to understand science as a new form of literacy for economic development. The election of Ronald Reagan in 1980 marked, for a time, a turn away from federal leadership in science education. Into the void stepped the American Association for the Advancement of Science with Project 2061, directed by F. James Rutherford, a participant in and witness to the inquiry-focused reforms of the previous period. Project 2061, which lasted throughout the reengagement of the federal government and the development of national standards in education, sought to expose students to a broader vision of science and the scientific enterprise with the same postwar goal of ensuring continued public support. That vision, however, became increasingly compromised by the shift to vocational goals and rote learning as a second wave of competition (this time from China and India) and concerns about the country's technical human capital in the new global economy buffeted the United States.

One of the central arguments of this book is that World War II marked a watershed moment in how the methods of science were presented to students and the purpose such understanding was thought to serve. In the last decades of the nineteenth century and those leading up to the war, the teaching of scientific method, whether advocated by scientists or by progressive science educators, was meant to convey habits of personal value, such as moral improvement or problem-solving skill. Following the war, curricular portrayals of scientific process aimed to promote an understanding of science and scientific work that would benefit the institution of science. That is, after the war, the goal was more about safeguarding scientists and the scientific enterprise as a whole. The reasons for this will become clear as I trace the changing role of secondary schooling in the United States, the shifting views of the value science offers society, and the rise of technological mastery as a central element in national advancement and global economic competition.

Ultimately, the importance of this historical account lies in what it reveals about the role that high school science has played in shaping the relationship between science and the public, in creating a particular set of conditions for how science is ultimately relied upon, supported, and governed by the people in the United States. Science, both as an abstract intellectual pursuit and as a publicly funded institution, has always existed at the margins of mainstream society, isolated for the most part, sometimes esteemed, other times reviled, oftentimes both. But it has long been deemed essential to the future of humankind. School science is the enduring, formal connection that exists between the scientific enterprise and the American people. Its role in attempting to mediate that relationship through teaching about the process of science is the story told in the chapters that follow.

1   From Textbook to Laboratory

DURING THE SUMMER OF 1893, nearly 30 million visitors made their way to the South Side of Chicago to tour the marvels of the World's Columbian Exposition. While many came to the fair to ride the great Ferris wheel anchored on the Midway Plaisance, take in the exotic Egyptian dancers, or ride the boats in the lagoons interspersed among the buildings of the Great White City (as it was called), others were intent on studying and learning from the hundreds of artistic, industrial, and cultural exhibits on display. Indeed, "the central idea of the Exposition," as one guidebook explained, "is to educate by making all displays an exponent of the world's advancement." This had been true of all such events held during the late nineteenth and early twentieth centuries. "Each succeeding World's fair, beginning with that held in London in 1851," visitors were told, "has been the schoolmaster of the nations." With the assemblage of displays and exhibits at this fair, Chicago was no exception.[1]

Formal education, in particular, occupied a prominent place in Chicago that summer. Exhibits of instructional methods and materials from public schools, technical institutes, and universities were splendidly arrayed in the Department of Liberal Arts, located at the south end of the Manufactures Building, an enormous, Corinthian-style structure—more than a third of a mile long and covering thirty-plus acres—that hugged the Lake Michigan shoreline. Befitting the importance of the fair's stated mission to educate the public, the location signaled "a position of high prominence in the center of interest and in the grandest of all the great structures."[2] The fair commissioners were in happy agreement on the importance of showcasing the nation's achievements in teaching. However, a close inspection of the displays themselves revealed less consensus about just which achievements reflected the best educational practices of the period.

The displays, in fact, seemed to capture a snapshot in time of an important transition then occurring in American teaching methods. Visitors strolling the aisles under the 212-foot ceiling could see, for example, all of the materials commonly found in traditional academic settings. There were elaborate textbook exhibits from the big publishing companies such as Ginn and Company and D. C. Heath alongside state-of-the-art desks arranged in precise rows courtesy of the United States School Furniture Company. These were the objects that came to mind when people thought of schools—young scholars arranged in rows, diligently poring over their books. Fairgoers could even hear wax recordings of recitations and oral examinations from students in five different states courtesy of Thomas Edison's recently invented phonograph. "Attendants are always ready to set the machines in motion and allow the visitors to hear a repetition of work that has actually been done in school," noted one enthusiastic observer.[3]

The books, desks, and recitations all represented the finest examples of traditional pedagogy. However, after turning a corner visitors encountered something altogether different, displays highlighting more radical methods of education. These included "exhibits of shop work, of decorative design," or of some other activities "in which tangible results have been secured." On view were drawings and clay work alongside manual-training products such as sewn garments, carvings, small tables, and kitchen shelves. Here was education of the eye and hand—"learning *by doing*." It was an approach to schooling far removed from the fact-based textbook study many then accepted as the standard of good instruction. Discerning visitors, a reporter insisted, "cannot fail to be impressed with one important fact—that the work of the new education is finding its way into schools of every sort."[4]

This shift away from book learning was readily apparent in the teaching of the natural sciences that had risen to prominence in the schools over the course of the nineteenth century. It was in subjects such as physics, chemistry, and botany that the new approach made its mark most strongly. Laboratory instruction, with its attendant student notebooks and technical apparatus, had taken the education world by storm by the time of the fair. "Several exhibits were designed especially to show the apparatus used for teaching the principles of science in secondary schools," noted one observer. Among the most prominent was that of the Harvard physics department, which in its development of a new laboratory course for high school physics had set the national standard for teaching by hand. The Harvard course, introduced in the late 1880s, had by the time of the fair achieved a measure of celebrity status. "Entering the [Harvard] exhibit from the south," reported one seemingly awestruck fairgoer, one found arranged "on five large tables . . . the apparatus that has made the 'Harvard experiments' in physics possible." Another observer praised the student laboratory work found in the Chicago exhibit as well, which "included pictures and descriptions of experiments, showing the material and apparatus used, the methods followed, and the results achieved." "These exhibits enabled the visitor to form a good idea of what the Chicago high schools attempt in science teaching, and how it is successful."[5] Laboratory teaching was without question the "latest thing."

The advancement of science and technology in the United States during the nineteenth century—which had produced wonders such as the wax cylinder recordings that filled the air with sound at the exposition, along with new industrial technologies that were transforming the national economy—had led to the embrace of the sciences in the schools. But just as these subjects settled into widespread acceptance in the curriculum, scientists and science educators began to change the focus of instruction from the information of science—the useful facts about the world found in textbooks—to the methods of its production. Laboratory work, in their eyes, offered much more than just the means to acquire useful knowledge. It had the power to promote mental discipline and provide moral uplift, the very things the long-established subjects of mathematics and classical languages were believed to provide. Casting the sciences as one of the liberal arts, however, came with a cost. Ironically, the more "hands-on" they made science teaching through laboratory work, the further removed it became from the practical and the everyday experiences of students.

•   •   •

The public embrace of science, both culturally and institutionally, came not all at once but rather in fits and starts throughout the nineteenth century. In the 1830s and 1840s, science made its way to the public by a number of routes. The public lectures and classes that began in New England and then spread to the Midwest as part of the lyceum movement devoted a significant amount of time to scientific topics. Mass-circulation newspapers and magazines were also prominent sources of information about science in that period. In 1845 the newspaper *Scientific American* first rolled off the presses in New York, offering readers descriptions of popular technologies of the day; it touted itself as an "advocate of industry" reporting on "scientific, mechanical and other improvements." Science was a topic of discussion in family parlors and social gatherings and could be found in children's books as well. Scientists themselves sought to advocate for their profession through the formation in 1847 of the American Association for the Advancement of Science (AAAS), the regular meetings of which were widely reported in the press. And the wonders of science and technology were further popularized, of course, by their celebration at public fairs and industrial exhibitions held in Philadelphia, Baltimore, New York, and other cities. Such public displays and events emerged as a regular part of the nineteenth-century American entertainment experience.[6]

As much attention as science garnered during the decades before the Civil War, it was surpassed by the surge in popularization that came in the years after. Newspapers and magazines, such as the *New York Times, Harper's Monthly,* and the *Atlantic,* began devoting more regular space to scientific topics in their pages. In 1872 the science writer E. L. Youmans founded *Popular Science Monthly,* a widely read periodical that fed worldview-changing ideas such as biological evolution, the mechanical theory of heat, and conservation of energy to an eager public. That same year, at the invitation of leading American scientists, the British natural philosopher John Tyndall (said to be "the greatest living popular scientific demonstrator" of his time) enthralled packed halls across the United States with wondrous lectures on heat, light, crystals, radiation, and other physical phenomena. The spectacles of Tyndall were followed in 1876 by the British biologist Thomas Huxley, who came to America and walked audiences through the evidence for Charles Darwin's theory of evolution, a topic that excited tremendous public fascination given the implications it had for humankind's place in nature, particularly following the publication of Darwin's *Descent of Man* in 1871.[7]

Operating alongside these popular outlets were the country's institutions of higher education, which offered formal instruction in various scientific fields throughout the nineteenth century. Given the widely accepted utilitarian framing of science at the time, the natural home for such teaching was the technical school or institute, which trained students for engineering or other specialized fields that supplied personnel to American industry. Schools such as Rensselaer Polytechnic Institute (founded in 1824), Yale's Sheffield School (1846), Harvard's Lawrence Scientific School (1847), the Massachusetts Institute of Technology (MIT) (1864), and later the Stevens Institute of Technology (1871) were all established with an eye toward training engineers and mechanics who could use science for material advancement. Science education—because of its instrumental value—was prominently featured in these institutions, which specialized in technical training over the kind of liberal education that was common in the regular colleges and universities of the time. In the middle years of the century, however, science advocates began seeking a place for science in the general college curriculum as well.[8]

One of the most notable—and unflinching—arguments for the broader incorporation of science in the curriculum was made by the British philosopher and naturalist Herbert Spencer in his 1859 essay "What Knowledge Is of Most Worth?" Railing against what he saw as the merely ornamental study of classical subjects such as Latin and Greek that predominated in colleges and universities, Spencer, in this "educational classic," insisted that knowledge of science was far more relevant than those dusty subjects of the past for meeting the immediate life needs of students and the public. Knowledge of scientific principles was of course necessary for the efficient pursuit of manufacturing and industrial concerns, he noted. But it had value for nearly every other life activity as well, from ensuring individual health and safety to raising children to promoting citizenship and aesthetic appreciation. The disciplinary knowledge of biology, chemistry, and mechanics, Spencer explained, was far more useful in daily affairs than was an understanding of verb conjugations or poetic myths in some dead language. Thus, to the question "What knowledge is of most worth?" he answered emphatically: "The uniform reply is—Science. This is the verdict on all counts."[9]

Spencer's utilitarian argument for teaching science found a welcoming audience in the United States. His essay, which expressed "probably the most thorough-going claims that have ever been made for the values of the study of the natural sciences," strongly supported the sentiments of those interested in bringing a new practical sensibility to higher education in America. Originally published in the *Westminster Review,* this "very able and common sense article" was republished almost immediately by the *New York Times.*

In a follow-up commentary a week later, the *Times* reasserted this claim to utility and pointed out that the embrace of science in higher education that Spencer called for had, in fact, already begun. "Twenty years ago, few of our Colleges had more than one or two men to deliver the lectures and guide the studies of pupils in the whole circle of natural sciences," they wrote. "All this has been changed, but so gradually, and only on the compulsion of a slowly-changing public sentiment, that we still hear the complaint that the Colleges are busy as ever teaching only the literature of a dead language, and ignoring the fact that this is the nineteenth century, with a living language and a live world enjoying its light."[10] The reign of science that Spencer predicted seemed to be coming to pass.

Only three years later, in 1862, Congress—at the urging of vocational-education partisans, merchants, and industrialists (and in the midst of the Civil War)—passed the Morrill Land-Grant Act, which provided for the establishment of universities that would "teach such branches of learning as are related to agriculture and the mechanic arts . . . in order to promote the liberal and practical education of the industrial classes." States quickly made use of the federal funds to establish new schools across the country, among which were Iowa and Kansas State Universities and the Universities of California, Illinois, and Nebraska. Previously established state universities, such as Rutgers (in New Jersey), Wisconsin, and Minnesota, were converted to land-grant status as well. All advanced the new approach to higher education that extended beyond mere technical or manual training, embracing the academic study of science as a foundation for its useful application to the broader affairs of American life.[11]

While the natural sciences were promoted primarily for the utilitarian advantages they held (for industrial application and for individuals in everyday life), a case was increasingly made in the years following the Civil War that such study had distinct liberal educational benefits as well. The argument came from a small group of scholars and scientists who saw science as an essential form of cultural education rather than just as a method of extracting greater efficiencies from steam engines or telegraph transmissions. In 1867 E. L. Youmans gathered up essays by leading science boosters such as Tyndall, Huxley, and physicist Michael Faraday and published them in a manifesto titled *The Culture Demanded by Modern Life,* the purpose of which was, in the words of one reviewer, "to vindicate for scientific and physical studies a prominent place in the system of liberal education against the predominance of classical studies." Not long after, Charles W. Eliot, a chemist by training who assumed the presidency of Harvard University in 1869, spearheaded the collegiate embrace of the natural sciences as legitimate subjects for liberal education. Although initially in favor of maintaining a distinction between scientific schools and liberal arts colleges, he succeeded in implementing the elective system of course selection, which was premised on the notion that both classical subjects (such as Latin and Greek) and scientific subjects had similar powers to train the intellect—a move that was key to breaking the hold classical studies had on the college curriculum.[12]

The willingness of leaders in higher education to see the academic value of practical studies both in established colleges (such as Harvard) and in the newer land-grant institutions soon led to the elevation of science, engineering, and agricultural coursework to the same status as literary studies. Indeed, by the early twentieth century, the presence of science in the collegiate curriculum was widely accepted. Looking back on the transformation, MIT president Richard Maclaurin commented on the impact the intensive public interest and support had made. It was in "the popular appreciation of science rather than in science itself," he claimed, "that the last century has proved absolutely revolutionary." Now the merits of science "are loudly proclaimed on every hand, and its importance is emphasized, with tired repetition, by college presidents and others."[13] Scientific subjects were no longer the poor stepchild of the college curriculum.

But exactly what sort of study was being advocated that would put the sciences on par with the classics? One line of thought had skill in the scientific method as the key learning outcome, at least for those of the elite classes. The astronomer Simon Newcomb—arguably "the most influential American scientist of the late nineteenth century"—strongly favored this aim. From his position first at the United States Naval Observatory and later as superintendent of the Navy's Almanac Office and president of the American Association for the Advancement of Science, Newcomb participated actively in public forums on the place of science and scientific research in American culture. In a special issue of the *North American Review* devoted to appraising the country's progress at its centennial in 1876, he lamented the poor productivity of American scientists compared to their European peers. Ultimately, science needed greater public support and, he argued, the public needed science—not for the knowledge it generated, but rather for its exacting rational method. The best way to secure this, in his view, was through education. "To fit the public for the increasingly complex duties it must undertake," Newcomb wrote, "I know nothing better . . . than a wide and liberal training in the scientific spirit and the scientific method." With "truth" as the primary aim of higher learning, there was no choice, he went on, but to let the scientific method be the "fundamental object in every scheme of a liberal education."[14]

As insistent as such calls were for greater attention to the process of science, few colleges made any real effort to have students actually experience scientific work prior to 1870. A handful of institutions had made the somewhat radical shift to laboratory teaching, "in which the student performs the experiments and uses the instruments himself," but such practices were hardly widespread. One commentator of the time noted that "the spirit of investigation showed itself actively enough in some directions, but it [did] not seem to have impressed itself upon the teaching." Another, looking back on the early days of science education in the state of Illinois, explained that the teaching in the normal schools there rarely involved anything remotely like scientific investigation. In a description replete with mocking imagery, he wrote: "Botany was chiefly a study of text," and chemistry "was apparently poured over the heads of the pupils like a shower bath." Physiology was "demonstrated to the imagination only," and physics was "taught as a department of mathematics, by deduction from first principles, with a sovereign contempt for apparatus and experiment, not merely implied but vigorously expressed."[15] In other words, it was about as far from hands-on study as possible.

• • •

The situation in the high schools through much of the 1800s was no better. Ministering as they did to a broader swath of the American people, high schools had always afforded a place for science in the curriculum. They were, in fact, established mainly to meet the needs of students seeking to fill the emerging commercial and professional occupations of a rapidly industrializing country, and the sciences, as Spencer later argued, were widely believed to provide the useful knowledge required for such work. Students typically mastered that knowledge through the methods of reading and recitation. Although teachers may have led off their lessons displaying an artifact for students to observe or interspersed their lectures with some well-planned whole-class demonstration, most of the instruction was little different from that found in the subjects of history and the languages—it consisted of teacher-led presentations and student rote memorization and recitation of the facts assembled in textbooks.[16]

In most cases this was by design. Responding to the increased demand in schools for "more practical" studies, William Torrey Harris in his position as St. Louis school superintendent put together a guidebook in 1871 titled *How to Teach Natural Science in Public Schools*. Aimed primarily at teachers of the lower grades, Harris's guide recommended either oral or textbook instruction and detailed the proper technique for conducting a recitation (little provision was made for direct student exposure to natural phenomena). The first step was for the teacher to "see that the main point is brought out, explained and illustrated again and again by the different pupils, each in his own language." "Every recitation," Harris explained, "should connect the lessons of to-day to the lessons already recited, and the questions awakened in to-day's lessons should be skilfully managed to arouse interest in the subject of to-morrow's lesson." The booklet by Harris, who later assumed the post of U.S. commissioner of education, proved so popular that it was published multiple times in the St. Louis schools' annual report and eventually reprinted as a stand-alone volume in 1895. Other school reports frequently singled out "recitations and the use of textbooks" as the most desirable methods of instruction as well.[17]

In practice, conducting a recitation was a fairly straightforward, though mind numbing, affair. Conway Macmillan, a professor from Minnesota, bemoaned the instructional "atrocities" of this approach "calmly perpetrated under the name of botany," and offered a typical example. "It is unimportant to locate definitely this recitation either in space or in time," he said, for the practice was so common that "doubtless it will be recognized all the way from Maine to California." The teacher begins:

> *Teacher:* "We have now to discuss the subject of cotyledons. What are the primitive seed-leaves called? All together now!"
> *Class:* "Cotyledons."
> *Teacher:* "Right! The cotyledon is the primitive seed-leaf. If there is only one what is the plant called? All together again!"
> *Part of the class:* "Monocotyledonous."
> *Some on the back seats:* "Polycotyledonous," "Polycoletedonous," "Mollycodledonous," etc.

Teacher: "Monocotyledonous is correct. Miss Smith, can you explain the
   derivation of the word?"
Miss Smith: "It is from the Greek and means one-cotyledoned."
Teacher: "But if there are two cotyledons what is the plant called?"
Miss Smith: "Dicotyledonous, meaning two-cotyledoned."
Teacher: "Also from the Greek?"
Miss Smith: "Yes, sir."
Teacher: "Very good: that is sufficient."[18]

Such interactions would continue in this fashion until the lesson was complete. The primary source of information was the textbook, which, wrote one observer, was "supposed to cover practically the whole science." The result for the student, as a critic remarked dismissively, was a "mere feat of memory of just as much value, perhaps, as the committing to memory of so many lines of 'Paradise Lost,' certainly no more."[19]

The emphasis on textbook-based science instruction became even more pronounced in the 1870s with the surge of public interest in science when several states enacted legislation requiring teacher proficiency in the sciences as a condition of obtaining their license. Written exams had become the new way to ensure mastery of the material. In Illinois, to take one example, repeated pronouncements by various state superintendents emphasizing the need for teaching the "practical knowledge of the sciences in their application to the ordinary pursuits of life" contributed to the Illinois legislature mandating in 1872 (without much warning) that all applicants for county teacher's licenses pass examinations in physiology, botany, zoology, and physics. This resulted in a science-teaching boom in the state teachers institutes, normal schools, and high schools to accommodate the new requirement. Teachers needed to know the facts, and know them quickly, in order to pass. "The natural sciences are upon us," one teacher exclaimed, "and we must do the best we can."[20]

The need to prepare more and more teachers for certification exams in the sciences (indeed, in nearly all subjects) during this period created a strong market for concise study guides. Publishers rushed to print textbooks and various pamphlets that could be used to help students in their test preparation. By far the most common of these books were those from the popular Fourteen Weeks series by J. Dorman Steele, a prolific author of textbooks during the boom years of school science expansion in the 1870s. Steele's books, published by A. S. Barnes and Company, were essentially cram books covering all the main tested subjects. Each was designed to impart the information contained as efficiently as possible, and it was all done via text. The detailed illustrations contained in the series were often as close to nature as students ever got. In the preface to his *Fourteen Weeks in Zoology,* Steele explained that this work was "prepared upon the same general plan as the preceding books of the Series." Among the principal features were "brevity," "directness of statement," "frequent foot-notes, containing anecdotes, curious facts, explanations, etc.," and "a uniform system of analysis in bold paragraph titles," all of which contributed to the goal of students securing passing grades on their examinations. Other publishers issued similar offerings, highlighting their test preparation power, and as more and more states implemented exam-based certification requirements, these volumes gained widespread popularity and became the default textbooks for use in high schools and normal schools across the country.[21]

In an attempt to characterize the state of science instruction following its rise in schools during these years, the United States Bureau of Education commissioned a national survey of chemistry and physics teaching, which affirmed the prevalence of textbook-based instruction in the high schools. Although the author, Frank W. Clarke, a Boston geochemist, noted the movement toward some individual student laboratory work in the colleges and universities, he acknowledged that "in the great majority of cases" in high schools, "mere text book work is done, [with] only a few experiments being performed by the teacher." Out of 607 schools surveyed, only four offered a full-year physics course with a student laboratory component. Nearly 67 percent of the year-long courses involved no experiments of any kind, "merely text-book recitations," he reported.[22]

Given the circumstances of the time, it would have been unrealistic to expect anything more. The recitation method was the dominant mode of instruction in the literary subjects throughout the 1800s, and the arrival of science in school classrooms did little to change that. As the sciences became more and more widespread, teachers increasingly found themselves overworked and teaching outside their areas of expertise, often culling information from the book not much ahead of their students. Critics charged that it was common for teachers to have the "text-book continually open" during recitations "in order to recognize if the answers are correct." One geology professor from Iowa told of encountering an excited schoolteacher at Yale's

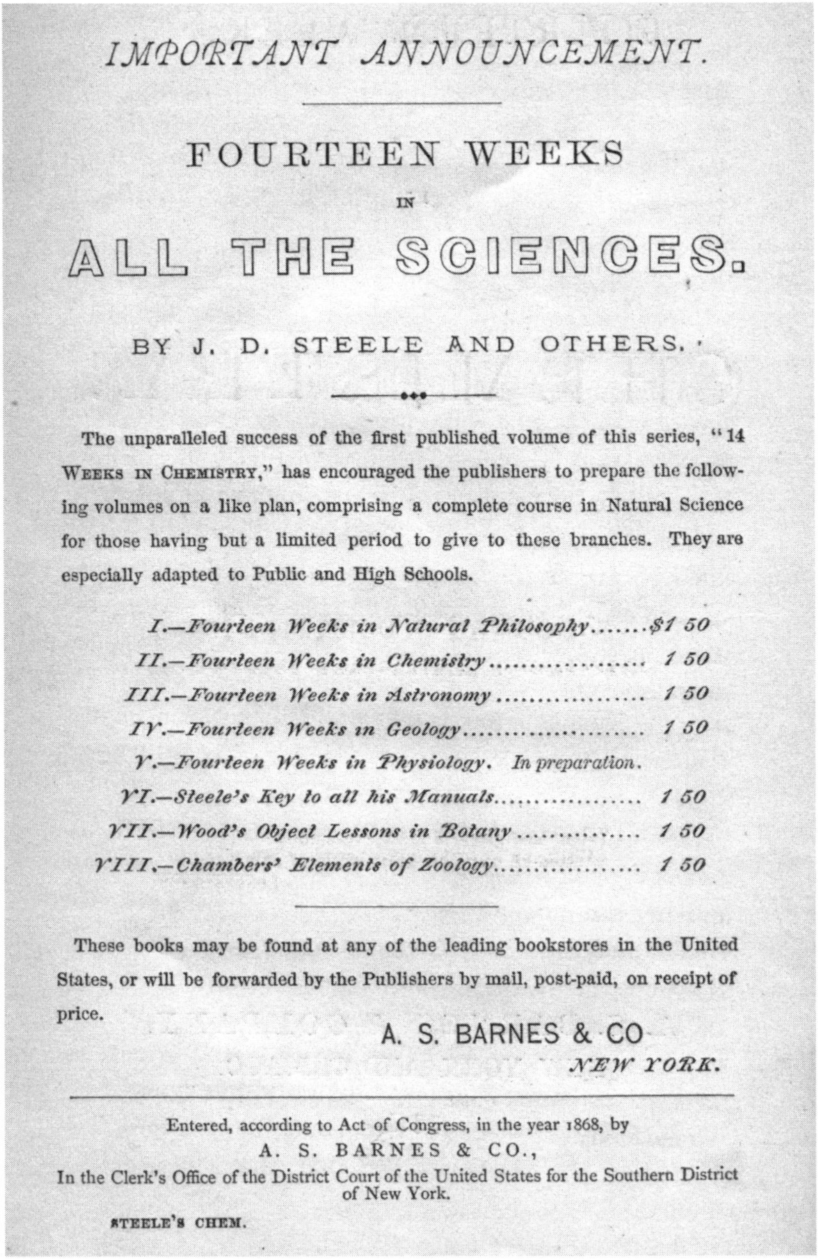

IMPORTANT ANNOUNCEMENT.

FOURTEEN WEEKS

IN

ALL THE SCIENCES.

BY J. D. STEELE AND OTHERS.

The unparalleled success of the first published volume of this series, "14 WEEKS IN CHEMISTRY," has encouraged the publishers to prepare the following volumes on a like plan, comprising a complete course in Natural Science for those having but a limited period to give to these branches. They are especially adapted to Public and High Schools.

I.—*Fourteen Weeks in Natural Philosophy*.......$1 50
II.—*Fourteen Weeks in Chemistry*.................... 1 50
III.—*Fourteen Weeks in Astronomy*................. 1 50
IV.—*Fourteen Weeks in Geology*...................... 1 50
V.—*Fourteen Weeks in Physiology. In preparation.*
VI.—*Steele's Key to all his Manuals*................. 1 50
VII.—*Wood's Object Lessons in Botany*.............. 1 50
VIII.—*Chambers' Elements of Zoology*................. 1 50

These books may be found at any of the leading bookstores in the United States, or will be forwarded by the Publishers by mail, post-paid, on receipt of price.

A. S. BARNES & CO

NEW YORK.

Entered, according to Act of Congress, in the year 1868, by
A. S. BARNES & CO.,
In the Clerk's Office of the District Court of the United States for the Southern District of New York.
STEELE'S CHEM.

Figure 1. Advertisement for books in the Fourteen Weeks series in J. Dorman Steele, *Fourteen Weeks in Chemistry* (New York: A. S. Barnes, 1871), 4. Reproduced from a copy in the University of Wisconsin Libraries.

Peabody Museum who had only ever encountered science in books. As he told it, she "was so anxious to handle the feldspars in one of the cases, because she had been teaching feldspar to her pupils for fourteen years and had never before seen the mineral."[23]

In addition to time constraints and lack of teacher expertise, another factor determining the nature of instruction was the public view of science as primarily an "information study." Despite claims from the likes of Newcomb regarding the intellectual value that came from understanding the process of science, it was clear that the utilitarian bias of the public maintained a strong hold on the high schools. "Men came to expect every conceivable good" from science, observed an early education historian. "For themselves and for others, they desired encyclopedic knowledge." The Illinois ecologist Stephen Forbes best captured the power of this expectation across America in a metaphorically rich observation he made in his 1889 history of science education in Illinois: "The beetle that drove the wedge home and struck the blow that split the log was really the *practical*." The sciences, he wrote, "were added to the public school course because it was hoped that a knowledge of them would help the people to live, and especially that the lot of the countryman and of the workmen in towns would be ameliorated if they knew more of the facts and laws of matter and of life."[24] There were few methods better tuned to the raw transmission of facts than rote memorization of textbook content.

●   ●   ●

The practical rationale on which science instruction made its way into school classrooms, however, proved to have a limited shelf life. Pressure to change the nature of high school teaching came quickly from science faculty at elite colleges and universities as they began to embrace European higher education practices, particularly the German model of advanced research and training. It was a model that emphasized research for its own sake, and in its American incarnation it exalted scientific study of a decidedly nonutilitarian character. Drawing on their own experiences abroad at places such as Göttingen, Berlin, and Munich, American scientists who came back from study abroad increasingly called for more time and better facilities to pursue this intellectual work. The research laboratory and teaching laboratory were fused in the German model, and laboratory instruction, made famous by the chemist Justus von Liebig in Giessen, quickly came to be the standard for sound science instruction in American higher education.[25] As important as the desire of the country's scientists to embrace this new professional research identity was the ease with which a pedagogy grounded in laboratory work aligned with the predominant moral aims of American schooling in the late nineteenth century.

Although some schools provided laboratories for students as early as the 1820s (Rensselaer Polytechnic allowed student laboratory work soon after its opening in 1824), the scale was typically small and largely focused on engineering applications. The recognition of the laboratory as a mode of general, or liberal, education came later with the establishment of teaching laboratories in the larger universities and liberal arts colleges. One of the first, according to reports of the time, was established in 1849 at Harvard by Eben Horsford, a graduate of Rensselaer who had gone on to study under Liebig. He was joined in his efforts by the chemist Josiah P. Cooke in 1850, who became a convert to Liebig's teaching methods and helped develop a regular system of laboratory instruction in chemistry at Harvard. In physics, MIT set the pace as the first site of laboratory teaching in the late 1860s under the direction of Edward C. Pickering, and it wasn't long before teaching labs were being set up in new facilities and makeshift spaces across the country. In 1873 the botanist Charles Bessey cordoned off a small room in the Old Main building at the state college in Ames, Iowa—the first of the land-grant colleges—and, after furnishing it with a table, microscope, and specimens to examine, tacked up on the door a cardboard sign that read "Botanical Laboratory." It was, he wrote (with some hyperbole perhaps), "the first time this term had been applied to any room in any college or university in the west."[26]

Nothing, though, signaled the complete integration of the laboratory into the mission of higher education as much as the establishment of Johns Hopkins University in Baltimore in 1876. Under the guiding hand of its first president, Daniel Coit Gilman (who previously served on the faculty at the Sheffield School), Hopkins became emblematic of teaching and research in the German tradition. For Gilman, the laboratory served as "an organizing vision sweeping across the entire range of college life." To head the chemistry department, he brought in Ira Remsen, who had spent years studying in Germany, first in Munich under Liebig (after his move there from Giessen) and then under other eminent chemists in Göttingen and Tübingen. In physics, Gilman hired Henry Rowland, a graduate of and later professor at

Rensselaer Polytechnic. Rowland had quickly garnered a national reputation for his experimental virtuosity and commitment to laboratory instruction. For scientists such as Remsen, Rowland, and others committed to the research ideal, true understanding could come only from students experiencing science firsthand. The learning experience should be, as one scientist wrote, "not unlike that by which [the] facts and principles [of science] were first established." From the founding of Johns Hopkins to the end of the century, "the tendency," wrote a Boston chemistry instructor, was "to make the student, from the very beginning, an *investigator.*"[27]

Historians of science have long recognized that scientific investigation in nineteenth-century America possessed its own unique set of epistemological norms and methodological commitments. In contrast to the more abstract, theoretical work characteristic of French and German scientists, American researchers were for the most part wedded to "fact-gathering," an approach grounded in the ideal espoused by Francis Bacon early in the seventeenth century. Bacon insisted in his landmark work *Novum Organon* that knowledge of the world could be built only through the inductive method, which entailed the painstaking accumulation, grouping, and classification of observable facts, from which generalized statements about nature could be made. Bacon's methodological prescriptions were the basis for subsequent British writings on scientific process, such as John Herschel's *Preliminary Discourse on the Study of Natural Philosophy* (1830), William Whewell's epic volumes *The History of the Inductive Sciences* and *The Philosophy of the Inductive Sciences* (published in 1837 and 1840 respectively), and John Stuart Mill's *System of Logic* (1843). The writing of these scholars, along with the examples set by Newton and Darwin (whose accomplishments were often held up as models of scientific reasoning), set the tone for work in the United States.[28]

Discussions of methodology (to say nothing of practice), however, did not always toe a strict Baconian line. In MIT biologist William T. Sedgwick's account, all the inductive sciences "begin with observations or experiments," but they then "advance to generalizations called hypotheses or theories; and finally proceed to overthrow or establish these generalizations by deducing from them their necessary consequences and rigorously testing them." Sedgwick clearly saw the value of deductive reasoning. Many scientists and philosophers, in fact, began to back away from strict Baconian formulations of scientific method following the dissemination of Darwin's work on evolution, which departed methodologically from the strict inductivism espoused by Bacon. Some scientists wishing to support Darwinian theory thus chose to modify their accounts of method—allowing for hypothesizing and deductions from those hypotheses—in order to ensure that their normative accounts of method and Darwin's practice weren't in conflict.[29]

Whatever the subtleties of the philosophical writings of the time, in everyday discourse, science was nearly always presented as primarily inductive—that is, as a process of building knowledge from the certain foundation of direct observation of nature. That was the message routinely conveyed to the students and the public. In a talk celebrating the opening of the Kent Chemical Laboratory at the University of Chicago in 1894, Ira Remsen made this point clearly. The scientific laboratory, he said, "has impressed upon the world the truth that in order to learn anything it will not suffice to stand aloof and speculate, . . . it is necessary to come into as close contact with that thing as possible." Let things "be handled, observed, drawn, described, impressed upon the mind through as many channels of sense as possible," insisted another proponent of laboratory study of the time. Indeed, the aversion to speculation and excessive theorizing was strong among American scientists of this era, who felt most comfortable with the patient accumulation of natural facts. As late as 1902, University of Michigan president James B. Angell publicly praised the scientific method of thinking, which he insisted was characterized by "exact induction from known facts, rather than the loose knowledge and theorizing that marked many subjects before the scientific method was in vogue."[30]

Apart from its intimate connection with truth, the inductive method—as experienced in the laboratory—was seen by scientists as educationally beneficial for a number of reasons. The training of observation was perhaps mentioned most frequently. "The student of natural science," stated Harvard president Charles Eliot in an 1884 address at Johns Hopkins, "scrutinizes, touches, weighs, measures, analyzes, dissects, and watches things." He explained that "by these exercises, [the student's] powers of observation and judgment are trained, and he acquires the precious habit of observing the appearances, transformations, and processes of nature." Observational skill in this method of instruction was tied closely to the ability to draw accurate conclusions. The Chicago botanist Charles Barnes explained that in addition to the "cultivation of the power of accurate observation," effective science teaching would "develop the capacity for determining exact data and deriving therefrom accurate results." Laboratory study, in other words, was the path to both "ocular accuracy" and "instrumental accuracy," which together resulted in "logical accuracy"—the necessary outcome of scientific study. Given these goals, science could "be studied only in the laboratory, a place of work where induction is supreme."[31] For its advocates, there seemed to be no limit to the cognitive benefits scientific study would bring.

•   •   •

Proponents of laboratory teaching frequently grouped these various aspects of learning (whether related to observation, precise thinking, or the ability to draw sound conclusions) into the more holistic notions of "mental discipline" and "training." These were often shorthand for the theory of faculty psychology prominent earlier in the century—the idea that the brain was a muscle consisting of parts or faculties, each dedicated to a particular skill or ability, and that these faculties could be developed through exercise (or would atrophy with neglect). In an address at the 1896 meeting of New York state science teachers, one participant stated plainly that the study of science indeed developed certain "faculties of the mind." Remember, he told the audience, that "the mind is stimulated to activity in certain directions by frequent exercise."[32] Even as late as 1906, Charles Eliot appeared to rely on the notion of mental faculties: "In the training of youth during the last fifty years," he commented, looking back on the rise of laboratory instruction, "the memory side has been developed" while "the observation side has suffered."[33] Much of this talk was undoubtedly meant to justify the inclusion of science in the curriculum alongside the classical languages, which had long been supported for the "discipline" their study provided.[34]

Although references to mental discipline and training were indeed common in educational discourse of the time, they represented less a consensus on psychological theory than they did old habits of speech intermingled with arguments from those clinging to the mental-discipline paradigm. For every proponent of faculty psychology, one could find a detractor. At the same 1896 New York meeting where the audience was reminded of how faculties might be properly exercised, another speaker claimed that "modern psychology has happily abandoned the hypothesis of faculties." Scientists and science educators just as often spoke of the ability laboratory work had to develop intellectual "power," which for some was synonymous with faculties of various sorts. "Every experiment performed, every phenomenon watched, every organism studied," wrote one New York teacher, "must do something towards developing power."[35] This notion of power was nothing more than the ability to reason intelligently, which for many advocates of laboratory work was one of the primary benefits students would gain from such study.

Whether one subscribed to the doctrine of faculty psychology or not, it was generally agreed that suffused throughout all of these presumed benefits of laboratory instruction was, perhaps, the most significant outcome of all—moral improvement. The laboratory experience was, in the eyes of scientists of this period, uniquely suited to accomplishing this goal. Nineteenth-century education had always had a strong moral component. "The schools call into constant activity the knowing, feeling, and willing powers of the child," and because of this, "all well-directed school work, therefore, is moral," noted one Chicago school leader. Charles Eliot explained that the subjects that made up the original liberal arts were chosen ultimately for their moral efficacy. "When we take the best intellectual and moral materials [of our day] to make up the list of subjects worthy to rank as liberal," he asked, "ought we to omit that natural science which in its outcome supplies some of the most important forces of modern civilization?"[36]

For Eliot the answer was obvious, and it was the laboratory in particular that gave science teaching its capacity for personal betterment. "Its fundamental purpose," he explained, "is to produce a mental and moral fibre which can carry weight, bear strain, and endure the hardest kinds of labor." Few were as direct in making this claim as the geologist William North Rice, who insisted in an address before the American Society of Naturalists in 1888 that "no one can become imbued in any measure with the spirit of science—the spirit of unselfish, courageous, reverent truth-seeking—without some degree of moral uplifting." He even suggested that studying the lives of scientists would show that "flagrant immorality has been exceedingly rare among scientific men—much rarer than among men of equal intellectual eminence devoted to literature, art, or almost any other pursuit."[37]

More specifically, the source of moral uplift could be traced to the essence of the inductive method itself, grounded as it was in a clear personal confrontation with the facts of nature. Johns Hopkins physicist Henry Rowland in an 1886 address commemorating the tenth anniversary of the university extolled this advantage of laboratory work. "Let the student be brought face to face with nature," he exclaimed, "the result is invariably humility, for he finds that nature has laws which must be discovered by labor and toil and not by wild flights of the imagination and scintillations of so-called genius." "Science teaches limits and bounds, the bounds of solid fact," the Iowa geologist William Harmon Norton went on in the same spirit. It teaches "fairness, emancipation from prejudice and personal bias, and a supreme love of the truth." Quite simply, "the study of objective things by means of the scientific method," many agreed, "not only trains the intellect but . . . tends to improve human conduct."[38] There seemed to be nothing the study of science couldn't accomplish.

Closely related to the cognitive and moral benefits that came from direct contact with nature was the value derived from the physical activity the laboratory required. For many, book learning was for the effete, and its passivity opened the door to indolence. Laboratory work, on the other hand, moved the pupil to action, and action was an unqualified good in late nineteenth-century American culture. Henry Rowland made this point clearly: "The object of education is not only to produce a man who *knows,* but one who *does;* who makes his mark in the struggle of life and succeeds well in whatever he undertakes." The laboratory, where students "stand face to face with nature," was essential, in his view. "They must try experiment after experiment and work problem after problem," he explained, "until they become men of action and not theory."[39]

Not only did laboratory-based science hold its own with the traditional liberal arts in its power to train the mind and instill morality, but its tendency toward activity embodied the best elements of manual training as well, a popular area of educational focus at the end of the nineteenth century that sought to bring the ennobling power of hand labor into the classroom. This was part of the new education on display at the Columbian Exposition. The successful introduction of laboratory work, in the words of William Sedgwick, has "justified the educational value of the methods of the gymnasium and the workshop." Another advocate of laboratory study echoed this sentiment, noting that in addition to providing "splendid mental growth," the laboratory sciences "furnish[ed] a kind of manual training of a high order of merit." One observer of the time even claimed that the shift from textbook-based instruction to laboratory work was "coming about largely as a result of the recent agitation [from] the manual training movement"—a movement that carved out its own niche in new shop courses and even separate manual-training high schools in cities across the country.[40]

Rank-and-file science teachers equally believed in the moral efficacy of laboratory work, whether it stemmed from its inductivism or the physical manipulation involved. Responding to calls from New York business and professional men in 1902 that schools focus their energy on "character building," "morals," and "manners," the vice president of the Eastern Association of Physics Teachers asked rhetorically at the organization's annual meeting: "What could assist in 'character building' more than hard and honest work in the laboratory at some difficult quantitative experiment, where the pupil must rely on his own efforts and depend for success upon truthful observation, accurate readings and a *strict following out of directions?*" (Laboratory work apparently didn't allow for much free play of the intellect.) Bemoaning the low numbers of those attending such courses in the high schools, he regretted that more students could not be touched by the benefits of such experience.[41]

Despite the repeated association of laboratory work with manual training, the movement toward the laboratory in the late 1800s took students further and further from the practical applications of science. Utilitarian justifications of the 1860s and 1870s were strikingly recast as the sciences gained admittance to what Eliot termed the "magic circle of the liberal arts." Or, perhaps more accurately, it was only through the shedding of their association with the practical that they were fully admitted to that circle. Supporters toward the end of this period could be found repeatedly minimizing the instrumental value of science in favor of its power to provide intellectual enrichment on par with the classics. If "scientific work is to be done for the purpose of attaining its chief end," exclaimed Kentucky professor James Lewis Howe, then "we must as far as possible lose sight of study for either practical ends, or for the purpose of general information; and we must as far as possible adopt laboratory instruction." The practical uses of science—the "beetle that drove the wedge home," in Forbes's words—had been effaced from the schoolroom.[42] Here was a great irony: initially brought into schools for the utilitarian knowledge and skills it offered the masses, scientific work was now valued primarily as a means to achieve abstract ideals of intellectual training and moral uplift rather than efficient and productive mastery over nature.

This new goal served a number of purposes in an era of social fragmentation in the years after the Civil War. As the professionalization of science gained steam in the later 1900s, depictions of scientific work as inherently virtuous contributed significantly to the social standing and political authority of scientists, allowing them to pursue research more for its own sake than for the practical contributions it might make. More important, though, science education was now aligned with the predominant views of late nineteenth-century education (which mixed in hard work, physical activity, and moral improvement with the development of intellectual power through the training of observation and reasoning), as laboratory study offered a particularly powerful way of preparing students for the duties of citizenship. In these years a virtuous citizenry—viewed as crucial to the flourishing of the fragile American republic—was the highest goal of public schooling, and practice in the process of science was uniquely suited to ensuring its production.[43]

2 The Laboratory in Practice

IN THE FALL OF 1893, the *Kansas City Times* ran a full-page feature on the city's "splendid new high school," then nearing completion. "When the doors are thrown open," the paper reported, community members, pupils, and teachers would find "a modern institution in all its equipment, outside and in." The school featured an observatory on the southwest corner "with a revolving copper dome, designed for astronomical purposes, which adds to the picturesque appearance of the building." Serious scientific work, though, was to be conducted in the many laboratory facilities, five in all, "each equipped with the latest scientific apparatus." The zoological laboratory was "well lighted and supplied with water and gas"; it was stocked as well with "microscopes for examining animaliculae" and with aquaria in which the "actions of aquatic life can be studied." The physical and chemical laboratories were fully outfitted too. The building was hailed as one of "the most modern and equipped schools of its kind in the entire West."[1] Kansas City would take a backseat to no one in its pursuit of the finest for the city's high school students.

Similar reports appeared in local newspapers across the country in the years around 1900. The *New York Times* in 1895 made sure readers knew of the fine appointments of Brooklyn Girls' High School on Nostrand Avenue—a "model twentieth century school building"—with its "three laboratories equipped with the best of modern appliances." The local editors in Chicago took time to express their pride in the biology laboratory to be built in Lake View High School on the north side of the city. "This is to be a model laboratory," they wrote. "There will be light from three sides of the room. The tables will extend in one line around the room so that each table shall stand in front of a window." The new high school in San Jose, California, was to have an entire wing dedicated to science. "On the ground floor will be the physiographic laboratory and class rooms, together with a private laboratory for the instructor," noted a reporter for the *San Jose Mercury* in 1907. "The electrical laboratory," the article further boasted, "will be far in advance of anything that has been built either on the coast or in the Eastern States."[2]

As secondary school enrollments exploded toward the end of the nineteenth century, local districts constructed new schools at a frenetic pace. In most communities, the high school building was the crown jewel of the educational system—the Girls' High School in Brooklyn, for example, "standing as it does at the head of the public school system . . . may properly be called Brooklyn's most creditable achievement," a reporter stated with pride.[3] In such circumstances, rare was the public official who was willing to neglect the features many accepted as essential to modern education, features that, more important, were often the markers of status sought out by civic boosters from one municipality to the next. The laboratory method of instruction had arrived, and no self-respecting school board wished to be caught without the facilities necessary to engage students in this important form of learning.

The dissemination of the laboratory method from the colleges and universities to a place of prominence in the high schools was facilitated by a number of activities and initiatives in the late nineteenth century. Proponents of such teaching, for one, wrote textbooks that incorporated laboratory work into classroom instruction. These new books, written beginning in the 1880s by practicing scientists trained in the European research ideal, began to supplant the more traditional, fact-based textbooks (such as the Fourteen Weeks series by Steele). In addition to new textbooks, scientists pushing for pedagogical reform began offering summer laboratory courses for struggling, poorly equipped teachers to train them in the new methods of instruction. Such initiatives originated at a variety of sites in the United States as well as overseas. In physics, the push came from Cambridge, Massachusetts, while in the biological sciences, key sites included Nebraska and even London, England. A willingness to abandon rote textbook learning in favor of laboratory instruction was in the air. The enthusiasm for this method rose to a point where nearly every course in the high school—science and non-science alike—saw the laboratory as central to student learning.

* * *

The path toward establishing laboratory teaching as standard practice in American high schools can be traced back at least to 1869, when Charles Eliot was tapped to lead Harvard University (such reforms often seemed to have their origins in the colleges). Although not inclined toward university-based research of the sort promoted at Johns Hopkins, Eliot had been, none-theless, an early and forceful advocate of laboratory teaching. In 1865, he took a position along with Francis Storer teaching chemistry at MIT and initiated student laboratory courses at the behest of the institute's president, William Barton Rogers. During their time together, Eliot and Storer drafted the first laboratory-based textbook in inorganic chemistry, which quickly be-came a classic in the field. When Eliot came to Harvard, he continued to push the bounds of pedagogical innovation. Laboratory teaching had already been established in chemistry under the guiding hand of Josiah Cooke, who had tutored Eliot as an undergraduate and first exposed him to the spark of experimental work. The physics curriculum, however, did not have the ben-efit of a figure with Cooke's interest or expertise, and so Eliot enlisted a young physicist by the name of Edwin Hall to fill that role.[4]

In March 1881, Eliot offered Hall the general physics lectures for freshmen and the responsibility for overseeing the physical laboratory. "For a person who can lecture and also give laboratory instruction," Eliot wrote to Hall, "the opportunity is an unusually good one."[5] Hall took over in September that year. In hiring Hall, Eliot reached into an extended network of scien-tists who were committed to laboratory teaching. Eliot, of course, knew all about building a program of laboratory instruction from his days at MIT, and he was well aware of the abilities and inclinations Hall would bring to his work at Harvard, for Hall had earned his PhD at Johns Hopkins under Henry Rowland, another strong proponent of the laboratory approach to sci-ence teaching.

In addition to Hall's teaching and laboratory responsibilities, he was given the task of administering the college's entrance exam in physics, which to that point was based solely on textbook study. He undertook this task duti-fully, if without enthusiasm. Few at the college seemed satisfied with the form or quality of the physics work typically submitted. In 1885, as part of a larger

Figure 2. Edwin A. Hall, ca. 1887. HUP Edwin H. Hall (1A), Harvard University Archives.

overhaul in admissions requirements at Harvard, a push was made to allow for laboratory preparation in high school physics as an alternative to the standard exam. As Hall remembered it, the requirements of such work were described only "in brief general terms, which left for the Physical Department the very considerable labor of arranging the details." A survey was made of the local high school physics teachers to gauge their receptiveness to student laboratory work and their capacity for implementation, and after some rather poor initial efforts, Hall was enlisted to develop guidelines for the new requirement following Cooke's lead in chemistry. The final version called for students to complete, as stated in the Harvard catalog, "a course of experiments . . . not less than forty in number, actually performed at school by the pupil." Hall listed the acceptable exercises, which were then included in a circular issued in 1886.[6]

Since 1883, Hall had been teaching a laboratory course on the principles and methods of physical measurement with John Trowbridge, a senior colleague in the department, and he drew on this experience in compiling the forty exercises that he deemed appropriate for students. Much of the work was taken from the more recent, laboratory-based textbooks of the period (such as Trowbridge's *New Physics,* Pickering's *Physical Manipulation,*

Worthington's *Physical Laboratory Practice,* and Gage's *Elements of Physics)* and covered topics in mechanics, heat, sound, light, magnetism, and electric currents.[7]

The quality of student work submitted for the first exam was uneven, to say the least. "The candidates," in Hall's appraisal, "came, like the traditional beggars, 'some in rags and some in shags and some in velvet gowns.'" The results prompted him to revise his original list to include not only detailed accounts of the exercises themselves but also specific descriptions of the necessary apparatus, where such apparatus could be purchased by budget-conscious departments, and how the laboratory work could best be integrated with existing textbook teaching. This version of the requirement—really more an entire course of work than an admissions guide—was published in 1889 as an eighty-three-page pamphlet titled the *Descriptive List of Elementary Physical Experiments Intended for Use in Preparing Students for Harvard College.*[8]

The tasks on the list were almost exclusively exercises in measurement, and although Hall counseled teachers that it would be unwise to expect "extreme accuracy" in the experimental results, he felt the teacher should "require of the pupil such accuracy as is fairly within his reach," otherwise the "laboratory work is likely to become a farce." In the first exercise, students were asked to calculate the breaking strength of a section of no. 27 spring-brass wire using a spring scale. From there they moved on to determine the compressibility of air using barometer tubing filled with mercury; calculate the number of vibrations of a tuning fork per minute by tracing the vibrating fork across a piece of smoked glass; measure electrical resistance in wires of different diameters and types using a two-fluid cell, galvanoscope, and commutator; and on through the rest of the forty assorted exercises. Hall and his colleagues believed such a course of experiments was best complemented with some lecture and discussion. But the bulk of the time— three forty-five-minute class periods per week over the course of a roughly forty-week academic year—was reserved for the preparation and completion of the exercises themselves.[9]

The laboratory notebook was the primary record of student work, and the physics faculty had high expectations for the form that record should take. In a letter that went out to area high school teachers after the June 1889 laboratory exam, Hall and his colleagues explained that "the record made during these exercises" should include "all the numerical data which the student has to obtain for himself, and all other observations of transient phenomena. These first notes must be in the identical book which is to be presented to the examiners." Moreover, the notes "should be written with a black pencil or in ink, and should not be altered after they are made." "Obvious, unquestionable errors occurring in them," they graciously allowed, "can be indicated and corrected in pencil or ink of a different color from that originally used without obscuring the original record." But they cautioned that such corrections "should occur rarely." "It should be made possible for the examiners to tell at a glance," they continued, "what notes were written at the time of the laboratory exercises and what were written at other times." The letter ended with the specific language the high school teacher should use to certify the authenticity of the notebook and the student's work therein, and all of this was to be submitted to the Harvard faculty for inspection at the time of the campus examination.[10]

The rigid nature of the exercises—their focus on precision measurement and the meticulous notebook write-ups—served to bring discipline to the mind and habits of the student, benefits that had been touted time and again by proponents of the inductive method toward the end of the century. Hall was definitely among the converts. "Physics as taught by the laboratory experience of the pupil," he acknowledged in an 1887 letter to *Science* announcing the new requirement, "gives a kind of training that is not given by any course of study *required* for admission to Harvard College or, perhaps, any other college in the country." This course of exercises, he noted, not only will "train the young student . . . to observe accurately, to attend strictly, and to think clearly" but also will "give practice in the methods by which physical facts and laws are discovered." In addition, such work will "make the pupil familiar with quantitative work" and, Hall explained, "give him a considerable amount of quantitative knowledge," which will add to the qualitative knowledge the student naturally acquired "by merely knocking about in the world." Such knowledge and skill, however, was not easily won; as Hall commented, echoing the existing associations of virtue, labor, and the laboratory, "Hard work is necessary for real discipline."[11]

The Harvard list's emphasis on facts and their precise determination was a natural outgrowth of the prevailing practices of Anglo-American scientific practice of the time. British science in the late nineteenth century, driven by industrialization and the development of the telegraph industry in particular, embraced research practices increasingly focused on perfecting

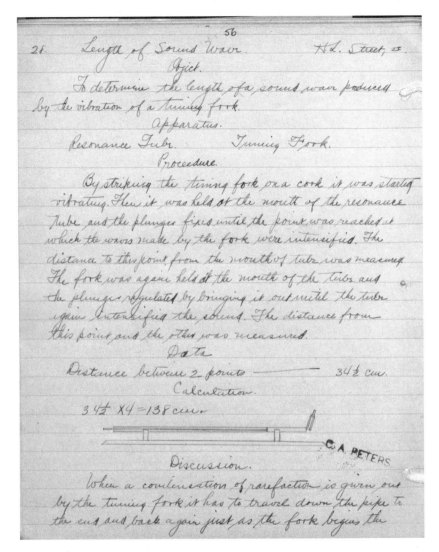

56

28. Length of Sound Wave. H.L. Street, □

Object.

To determine the length of a sound wave produced by the vibration of a tuning fork.

Apparatus.

Resonance Tube.   Tuning Fork.

Procedure.

By striking the tuning fork on a cork it was started vibrating. Then it was held at the mouth of the resonance Tube and the plunger fixed until the point was reached at which the waves made by the fork were intensified. The distance to this point from the mouth of tube was measured. The fork was again held at the mouth of the tube and the plunger regulated by bringing it out until the tube again intensified the sound. The distance from this point and the other was measured.

Data

Distance between 2 points ———— 34½ cm.

Calculation.

34½ × 4 = 138 cm.

C. A. PETERS

Discussion.

When a condensation of rarefaction is given out by the tuning fork it has to travel down the pipe to the end and back again just as the fork begins the

Figure 3. Page from a "Length of Sound Wave" experiment from the notebook of Harvey Louis Street II, Polytechnic Institute Preparatory School, Brooklyn, New York, ca. 1909, 56. Laboratory Notebooks / SC MS 0032, Special Collections, Lehigh University Libraries, Bethlehem, Pennsylvania.

methods and techniques of precision measurement necessary for industrial standardization. This physics of measurement fit comfortably with the empiricist research ethos of American scientists, who also focused their efforts on the accumulation of data and the perfection of instrumental techniques. In a convocation address on the state of physics in 1894, University of Chicago physicist Albert Michelson emphasized the importance of the field's focus on precision. "It seems probable," he stated, "that most of the grand underlying principles have now been firmly established and that further advances are to be sought chiefly in the rigorous application of these principles. . . . It is here that the science of measurement shows its importance." "The future truths of physical science," he added, citing an eminent physicist of the time, "are to be looked for in the sixth place of decimals."[12]

Indeed, American scientists demonstrated a flair for work of this nature. The country's first two Nobel prizes in science were awarded to Michelson, who received the 1907 physics prize for his work with precision optical instruments, and to T. W. Richards of Harvard, who garnered the 1914 chemistry prize for the accurate determination of atomic weights. The prizewinning research of both men was conducted during the 1880s and 1890s. Henry Rowland's work at Johns Hopkins was in the same mode: it centered on the manufacture and use of high-quality diffraction gratings, tools that a colleague called "beautifully simple instruments of precision which have contributed so much to the advance of physical science." And Edwin Hall himself was no exception, having made his name in physics before his arrival at Harvard for the discovery and precise determination of the influence of magnetic fields on electron-flow distributions in current-carrying conductors, a phenomenon that became known as the Hall effect.[13]

As the expectations for quantitative laboratory study were formalized, Hall commented on the concern some expressed that such work would prove to be too demanding for schools to implement. The main obstacles were believed to be the cost of laboratory facilities and apparatus and the lack of teacher expertise in overseeing such work. Although outfitting schools with scientific apparatus was no doubt expensive, the trend toward laboratory work encouraged the development of inexpensive materials for use in schools. One Massachusetts physics teacher, Alfred Gage, went so far as to develop his own line of laboratory equipment, operating his business out of Boston English High School, where he taught. Gage's apparatus, however, was designed primarily for the illustration of qualitative phenomena in physics and,

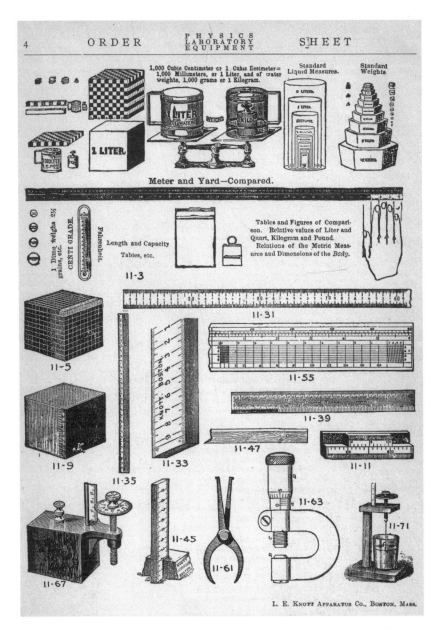

Figure 4. Apparatus for the National Physics Course from *1907–8 Order List of Physics Laboratory Equipment,* L. E. Knott Apparatus Company, Boston, 1907, 4. Reproduced from a copy in the Cornell University Libraries.

as such, failed to meet the requirements of the Harvard course. This led Hall along with Wallace Sabine, a departmental assistant at Harvard, to design their own low-cost apparatus for use with their specified quantitative exercises. This was the apparatus destined for the display tables at the 1893 Chicago World's Fair.[14]

The issue of teacher expertise was less easily addressed. Of the scientists who took an interest in the local schools, many had rather low opinions of the quality of classroom personnel. "Were it merely demanded that the instructor should hear the students repeat by rote the statements in some trashy 'Fourteen Weeks' text book, there would be little trouble," writers in the *American Naturalist* editorialized. "But experimental work demands more training and more brains." To provide this training, Harvard established a summer school course for area teachers. The course, first offered in 1887, enrolled eighteen teachers by the summer of 1889, and in 1890 a second, more advanced course was given. The experience gained from the summer school and the production of the new quantitative student apparatus was the basis for the 1891 publication of a physics textbook jointly authored by Hall and Joseph Bergen (a Boston high school teacher and the instructor of Harvard's summer course for teachers) that provided a fully integrated manual of laboratory physics that met the Harvard requirement. In many ways, the textbook was just a revised and expanded version of the original Harvard pamphlet.[15]

While the efforts that Hall and others made to ease the transition from traditional textbook teaching to laboratory instruction were well intentioned and considerable, in the end it was the presence of the Harvard requirement itself that was primarily responsible for the shift in high school teaching. Indeed, Eliot's biographer noted that Eliot deliberately used "the definition of college-entrance requirements" as "a lever to raise the schools"; that is, the entrance requirements occupied "a point of what might be called high tactical value in the campaign for [high] school reform." Hall seemed to acknowledge this in a comment he made in his letter to *Science* announcing the requirement. He admitted that the move to laboratory teaching would require considerable effort on the part of the schools. "Just how great the difficulties which have been touched upon will appear to the preparatory schools," he stated, he was "unable to foresee, but there can be little doubt that the larger schools which send boys to Harvard will, in general, speedily adopt the experiment method for preparing boys in physics." And adopt it they did.[16]

• • •

The situation in the biological sciences was not nearly as well developed as it was in the physical sciences. Although educators in the 1880s, particularly in the Boston metropolitan area, were inspired by Harvard's eminent zoologist Louis Agassiz and his famous entreaty to "study nature not books," his public lectures and summer school courses had more influence on schooling in the lower grades than they did ultimately on work at the high school level. The advancement of laboratory study in biology (an emerging term for the collective fields of botany, zoology, and physiology) in secondary schools owed more to the life scientists who came later.[17] In biology, there was no center of reform comparable to Harvard for physics. There were, nonetheless, pockets of activity that began to offer models of desirable high school laboratory work that schools soon began to emulate.

One of the more notable regional centers of reform was in Nebraska, where the botanist Charles Bessey worked to spread the use of teaching laboratories. Bessey, who had opened the first student botanical laboratory at Iowa State College in Ames before moving to the University of Nebraska in 1884, was a passionate advocate of the laboratory method as practiced in many of the German universities. From his faculty position at Nebraska, he worked with teachers throughout the Midwest and Great Plains states to encourage its dissemination to the nation's schools. Having spent some time himself as an elementary school teacher in Ohio early in his career, he was familiar with the needs of teachers and did all he could to help them from his position at the university. He provided them with laboratory supplies he obtained through the university at cost along with herbarium specimens he offered on loan for those who needed them, and he often taught summer courses for teachers interested in using laboratory methods in their schools—courses similar to those offered at Harvard for physics and chemistry teachers.[18]

His greatest influence, however, came through the introductory textbooks he wrote that dramatically reorganized botanical study around laboratory and field work. The first was his *Botany for High Schools and Colleges,* published in 1880, which was followed four years later by *The Essentials of Botany,* a modified version of the earlier book. Both volumes represented, in the words of one reviewer, "a new departure in American botanical textbooks" and went through multiple editions by the end of the century. "It is principally due to Charles Bessey," insisted one of his contemporaries, "that the methods of instruction in botany have improved so greatly."[19]

Much further east, across the Atlantic, teaching in the biological sciences was being shaped by the work of the renowned British zoologist Thomas Henry Huxley. Well known as a tenacious defender of Charles Darwin's evolutionary ideas, Huxley maintained a keen interest in promoting public understanding of science, particularly as it became apparent to him that a broad-based cultural acceptance of science was essential to its survival as a profession. In 1876 Huxley wrote *A Course of Practical Instruction in Elementary Biology* with the help of Cambridge-trained physiologist Henry Newell Martin. With material drawn from a short course of lectures given to teachers in London in 1872, Huxley sought to articulate a guide for the practical examination of specimens so that others might benefit from their close study. Martin, who moved from London to set up the biological laboratories at Johns Hopkins the same year of the book's publication (and of Huxley's American tour), was primarily responsible for the detailed laboratory directions it contained, which were praised for their precision and clear step-by-step sequence. Huxley and Martin took a bold stand on the importance of laboratory work (along with their unified conception of the life sciences as the singular "biology"), which spawned a raft of new American biology textbooks that modeled themselves on Huxley's laboratory approach.[20]

The American biologists recognized that they had come late to the game. Standards of student laboratory work in both chemistry and physics were well established by the late 1880s. Even with the efforts of reformers such as Bessey, Huxley, and Martin to promote the newer methods in biology, there was more work to be done. "No one is now satisfied to chase a plant into a corner by means of an analytical key and then clap a name upon it like an extinguisher," wrote one botanist at the time. The life sciences needed to move beyond the long-standing, relatively passive practices of identification and classification. "To learn chemistry," the botanist observed, "it is necessary to handle the re-agents, cut the fingers with test tubes, burn them with hot rods, stain them with corrosive acids, blow one's self up with fulminates, asphyxiate one's self with nameless gases." It was clear to him that "it is the same with vegetable forms: we must handle the plants, familiarize ourselves with their gross and minute anatomy, compare them with others, note how they are influenced by environment," and so on. The zoologists of the time were similarly inclined. An Illinois anatomy professor lectured that "teachers of zoology cannot do better than to accept the principles arrived at, and build their particular methods up on the well-earned experience of their brothers in the physical sciences."[21] The move to the laboratory in biology was on.

While biologists were intent on following the lead of the physicists and chemists in their emphasis on laboratory work, the nature of practical work with plant and animal specimens was notably different from what students did with reagents, wires, and springs—and by no means uniform from one biology instructor to the next. Rather than emphasizing mathematical calculation and precision measurement, as was common in the physical and chemical laboratories, work in the biological sciences assumed a more descriptive, qualitative cast. As fields of research, botany and zoology in this period were primarily sciences of identification and classification, and they remained so for some time despite the introduction of more detailed laboratory work.

Individuals working in these areas adhered to a kind of work in which detailed observation and the location of specimens within classificatory systems were the primary tasks of knowledge production, tasks that were well suited to the strongly empirical notions of science common among American researchers. Biologists, in other words, made sense of the diversity of living organisms visually rather than mathematically, and this visual study was the essence of botany and zoology, subjects that fell into the overarching category of natural history. Edwin Hall himself noted this fundamental difference between the physical sciences and natural history. "Pure observation of numerous, minute, varied details," he noted, "plays a much more important part in natural history than in physics." Taking the opportunity to poke fun, he related the extreme case of Agassiz, whose approach with a new pupil, he said, was "to set him to gaze in solitude at a single fish for two or three days."[22]

This emphasis on the visual carried over naturally into school exercises. Laboratory exercises most often involved having students prepare materials for examination, often under the microscope. Bessey himself was a strong proponent of microscope study. He argued for a rigorous analysis of plant anatomy—the kind of work that could only be done in the laboratory. His botany textbook of 1880 was designed for "constant use in the Laboratory, the text supplying the outline sketch, which may be filled up by each student, with the aid of the scalpel and compound microscope." His *Essentials of Botany* included a large number of exercises on various representative plant types. Students were instructed, for example, to "mount in alcohol a thin slice of ripe pea [and note] the small granules (along with the large starch grains) in the cells," to "make thin cross-sections of an apple leaf and note the intercellular spaces of the lower half of the section," and to "mount portions of [pond scum] and search for desmids [a type of green algae], using a 1/2-inch objective." The guiding principle in nearly all these textbooks was the steadfast belief that "the young botanist should not be content to obtain all his facts at second hand; he must see with his own eyes all that may be seen."[23]

The laboratory work outlined in the zoology and biology books that followed Huxley and Martin's lead was similarly focused on the systematic analysis of type specimens from key biological groups, which was designed to demonstrate the evolutionary relationships between organisms. A typical exercise in Charles W. Dodge's 1894 *Introduction to Elementary Practical Biology,* a popular textbook modeled on the Huxley approach, called for the examination of a starfish. It opens by surveying the variety of ways the specimens might be prepared, including dropping them live into vessels of alcohol, drying some in the sun, or soaking them in a solution of potash (a potassium-containing salt) followed by brushing away the macerated tissue to expose the skeleton. Dodge then gives instructions on how to prepare the vascular system for observation using an injection of carmine or Prussian blue. From such prepared specimens, students were asked to make drawings of the various surfaces (ventral and dorsal). Students observed and drew, in turn, the ambulacral feet, the radial water tube, the pedicellariae, and the various components of the skeleton, among the many anatomical features. Throughout the book, young scholars were busied with the tasks of preparing, describing, and drawing, a process they would cycle through again and again as they moved from one organism to the next.[24]

As the details of Dodge's textbook suggest, the recording of observations in the form of drawings was considered central to the learning process. The value of this pedagogical technique derived from its grounding in the epistemology of observation (as well as from its connection to what was then believed to be an essential element of manual skill). "The value of drawing, in giving directness to observation," the Illinois zoologist William Locy explained, "is recognized by all teachers." "These laboratory sketches," he elaborated, "should be viewed, not as artistic efforts, but as a means of expressing observations and conclusions in lines." It was, in other words, a form of "visual thinking." The work of making a drawing directly from a physical specimen in the laboratory was so important that Dodge purged his textbook of all illustrations to eliminate the temptation for students to copy from its pages.[25]

Although making drawings from real organisms was undoubtedly more engaging than memorizing species names and the taxonomic categories to which they belonged, there was little else within the exercises to pique the interest of high school students. In this move away from the practical, the biologists had much in common with the physicists and chemists. Botanists during this period were doing their best to distance themselves from the association of plant study with herbalism and applied agriculture. In 1892 a professor from Minnesota wrote that he looked forward to the day when formal laboratory analysis would be the norm in schools, replacing the need to study plants for utilitarian purposes. "When that day has come we shall see the end of the herbalist-regime," he foretold. "Botany will be yet more emancipated from the realm of the so-called 'practical' . . . and will be free to take its place . . . with its kindred sciences—geology, chemistry, physics— in the purer region where there remains something of potentiality."[26]

* * *

By the 1890s, many organizations and institutions moved to ensure the general spread of laboratory teaching in schools across the United States. It was no coincidence that Harvard president Charles Eliot had a prominent hand in shaping the policies and actions of these national organizations. But there was a local embrace of the new approach to science teaching as well that manifested itself in the inclusion of laboratory facilities in the new school buildings that were being constructed to accommodate the growing number of students enrolling in the nation's schools. If the top-down policies were the lever of change (as Eliot expected they might be), the grassroots community support in the form of civic pride proved to lighten the load in shifting classroom practices in the direction the reformers desired.

The most historically prominent statement in favor of laboratory teaching during this period came from the influential report by the Committee on Secondary School Studies, more widely known as the "Committee of Ten." The committee was formed under the auspices of the National Education

Association (NEA) in July 1892 and was chaired by Eliot. The ostensible purpose of the group was to promote greater uniformity in the country's high school courses of study so that they might be better aligned with typical college admissions requirements, to improve what many referred to as the "articulation" between the high schools and colleges. Decisions about uniformity, of course, could not be made without wading into the disputes between curricular traditionalists (who were wedded to classical studies) and the so-called modernists (who called for the inclusion of the sciences and modern languages in the new standards). With Eliot in charge, though, the manner in which these disputes would be resolved was hardly in doubt.[27]

At their first gathering in Saratoga, New York, in the fall, Eliot and the rest of the committee directed that nine subcommittees be organized around the various school subjects. There were three that dealt with the sciences: one devoted to a "Physics, Astronomy, and Chemistry" cluster, a second on "Natural History (Biology, including Botany, Zoölogy, and Physiology)," and the last addressing "Geography (Physical Geography, Geology, and Meteorology)." All the science subcommittees began their individual conferences on the same day later that year (December 28) in the greater Chicago area. The physical and biological sciences groups met on the campus of the University of Chicago, while the noted educational reformer and subcommittee member Francis W. Parker from the Cook County Normal School hosted the geography group in nearby Englewood. Each of the subcommittees was charged with describing, among other things, when the subject in question should first be introduced in the schools, how often it should be studied, and whether it should be a requirement for college admission.[28]

Members of the physical science subcommittee, which was headed by Johns Hopkins chemist Ira Remsen and included noted textbook author and laboratory advocate Alfred P. Gage, pushed hard for a greater commitment to laboratory work as "the best method of teaching" because of all the mental and moral benefits it provided. They recommended, among other things, that at least half the class time in physics and chemistry be devoted to lab work, that it be primarily quantitative, and that such work should be required for admission to college. "It requires no argument," Remsen wrote, "to show that the study of a text-book of Chemistry or of Physics without laboratory work cannot give a satisfactory knowledge of these subjects, and cannot furnish scientific training." To further encourage such work, the group offered up a list of fifty experiments in physics and one hundred in chemistry to be used as the basis for what it deemed "appropriate" high school work. These lists followed the format Hall had popularized a few years earlier, and in fact were vetted by Hall prior to their publication. In a private communication to Eliot, Hall conveyed his approval—mostly because the committee list was "as like the Harvard list as it could be made without exciting bitter opposition from certain authors and publishers . . . who do not like to see the influence of Harvard extend itself."[29]

The biology subcommittee was chaired by Washington, DC, school superintendent W. B. Powell and counted Charles Bessey among its members. Those in this group were even more adamant than the physical scientists regarding the importance of laboratory exercises. They recommended that students spend three-fifths of their classroom time engaged in such work over the course of a year. Whether the subject of study was botany or zoology (and the committee members were somewhat split on this question, though most recommended botany for high school work), the members wrote that the study, "to be of much value, *must consist largely of laboratory work,* actual work, by the pupils, with the plants or animals. This cannot be too strongly emphasized." And, like the physical science group, the biologists recommended that natural history be required for college and that the entrance examination include a laboratory component. The geography subcommittee, though it dealt with a subject that was rarely pursued in a laboratory setting, lobbied for such work just as much as the other two. The members emphasized the importance of having students learn in a way that paralleled the development of scientific knowledge. All geographic study, they agreed, should proceed following the inductive method; it should "develop the power and habit of geographic observation" and "arouse a spirit of inquiry."[30]

In his summary of the subcommittee reports, Eliot highlighted these calls for more systematic laboratory instruction. "All the Conferences on scientific subjects," he observed, "dwell on laboratory work by the pupils as the best means of instruction, and on the great utility of the genuine laboratory note-book."[31] Since such study required considerably more time than traditional recitation methods, Eliot suggested that schools go so far as to hold special Saturday morning sessions devoted to laboratory work in the schools, with the most senior students (who, at least in Eliot's mind, no doubt would be happy to spend their Saturdays at school) serving as lab aides to the teachers. As further evidence of the central place of the laboratory in the report's recommendations, the committee included "Laboratory and Field

Work" as a separate heading in its analytical index (with nineteen separate entries), thus providing a mean of easy access to all the various comments and recommendations related to the topic throughout the text.

The committee's recommendations, published in December 1893, drew considerable attention both in the popular press as well as in more specialized education journals. Many commented in particular on the "stress . . . laid upon laboratory work" in the document. The influence of the report as a whole, one observer noted, was "widespread and pervasive." Another recalled that "it took things by storm." Although some critics objected to the autocratic nature of the report, deeming it an "educational encyclical," before long schools across the country began aligning their pedagogical practices with its recommendations. One educator noted years later that the report "established precedents which have largely determined the nature of science instruction ever since."[32]

But as much as the Committee of Ten report put laboratory work at the center of public discussions of school science teaching, the real catalyst for classroom change came from the standardization of college entrance requirements. Movement in this direction came not long after the Committee of Ten disbanded. In 1895, the NEA assembled a second group of educational leaders to address not just the high school course of study in general but also the specific question of what an appropriate course of study for college admission should look like. The group was chaired by Chicago high school superintendent Augustus Nightingale, who delivered his report in 1899 after four years of work. The science subcommittees of this second group repeated many of the recommendations of the Committee of Ten, calling again for individual student laboratory work, including quantitative exercises to be written up in a laboratory notebook where appropriate. The subcommittee on physics was chaired this time by Edwin Hall, who, much to Nightingale's frustration, did little more than include a modified list of the Harvard experiments as his submission to the group.[33]

Nightingale's annoyance with Hall notwithstanding, the efforts of these committees seemed to be having the desired effect. It was primarily because of Harvard's initiative in physics that colleges were moving to greater inclusion of laboratory-based courses in their admissions requirements. This appeared to be obvious to Hall, who noted the year after the publication of the Committee of Ten report that Harvard's "examinations in experimental physics have served the very important purpose of *creating* a standard of work for the schools." By 1896, as the work of the second NEA committee got under way, nearly half of the country's most notable institutions of higher education included laboratory work in their guidelines for applicants. And of all the science subjects that made it onto the required list, physics outpaced the others by far.[34]

Some felt that recommendations from national committees, no matter how esteemed their membership, could do only so much to bring about change. And so, with the support of Eliot and soon-to-be Columbia University president Nicholas Murray Butler, the members of the Association of Colleges and Preparatory Schools of the Middle States and Maryland (one of many regional associations of the time) approved a resolution calling for the establishment of the College Entrance Examination Board (CEEB), which would offer standardized tests in nine school subjects for colleges to consider in admission decisions. The CEEB's role would be to formalize through written exams the curricular suggestions of the NEA committees. The first of these were offered in the summer of 1901, and those in the subjects of botany, chemistry, geography, and physics all required students to submit laboratory notebooks from their high school course in addition to sitting for a ninety-minute written test. The notebooks counted for anywhere from 30 to 40 percent of the total exam score depending on the subject. In the second year of operation, 443 notebooks were submitted from students across the country, a number that rose to 700 by 1906.[35]

Although the examination system itself was new, the standards the board set for high-quality work hewed closely to those previously established. In physics, for example, the "Definition of Requirements" included a list of experiments that was, as Hall pointed out a few years later, "precisely the same as that . . . in the report of the National Education Association and in the Harvard 'Descriptive List.'" Hall's influence, indeed, had been significant during these years, and the adoption of Harvard's requirement for experimental work as the de facto national standard moved high schools to adjust their offerings to meet those expectations. In explaining the change in physics teaching from the early 1880s, when there was "practically no individual laboratory work," to 1909, when schools had "laboratory work for everybody," University of Chicago physicist Charles Riborg Mann (a leading figure in the reform movement to come) pointed to the role played by the colleges and Hall. "The college entrance requirements have been of the greatest assistance in hastening this progress," he said, and added, "physics teaching

owes a great debt of gratitude . . . to Professor Hall and Harvard in particular, for this acquisition of laboratories for physical science in the schools." [36]

•   •   •

Initially urged on by scientists and scholars sitting at the nodes of power in
higher education, laboratory instruction rapidly gained a pedagogical and
cultural cachet that carried it deeply into local communities across the United
States. City school districts and rural communities invested heavily in new
schools with up-to-date laboratory space for all the science subjects, as was
the case with Kansas City Central High School in 1893. Others added science wings onto existing buildings or retrofitted classrooms to accommodate the new teaching approach, as was done at Lake View High School in
Chicago two years later. The wave of excitement for laboratory work soon
extended beyond the confines of the sciences as other subjects sought to appropriate the novel hands-on approach. Advocates of nearly any subject from
history to mathematics could be found insisting on some sort of laboratory-
based instruction for their students, and by the mid-1900s it appeared to
have fully saturated the high school experience of American students.

The construction of Lake View High School's new laboratory was but one
instance in the larger transition from textbook to laboratory teaching that
took place in the Chicago public schools in this era. An examination of the
district's financial reports from 1880 to 1902 shows a noticeable shift toward
practical work. In the early 1880s expenditures for laboratory equipment
were meager and routinely went only for the repair of existing apparatus or
"philosophical equipment," as it was called, and to purchase chemicals, most
likely for teacher demonstrations rather than individual student work. During
these years, as the Chicago schools' annual budget approached $1 million,
money allocated to this category hovered around $50 per year. The pattern
changed in 1885 with a one-time $4,100 allocation for the purchase of new
laboratory equipment, which was followed by a steady increase in yearly
spending on chemical and philosophical apparatus. More dramatic increases
began in 1893, the year of the Columbian Exposition and the Committee
of Ten report. And from then until 1900, the district spent nearly twenty
times more than it had previously to support laboratory teaching in its schools
during a period when the district budget as a whole increased only sixfold. [37]

As the newspaper account of the facilities at Lake View High School suggest, more than the routine allocation of funds was needed to fully imple-

ment laboratory instruction. In addition to these annual expenses for apparatus and supplies, the Chicago Board of Education approved thousands of dollars each year beginning in 1891 to retrofit existing high school buildings throughout the city with new labs, expending more than $24,000 at the peak of their efforts, from 1895 through 1896. The world had entered a new era. "That meager intellectual equipment which opened the ways to success in life a century ago, is all insufficient now," wrote the Chicago Committee on High Schools. "The difference in the demand for mind training now and then . . . is as great as the difference in the methods of locomotion and transportation now and then." And school facilities needed to reflect that. Explicitly citing the recommendations of the Committee of Ten, the board insisted on placing laboratories at the top of the list. "The days of the old text-book regime . . . are past," the director of Chicago's high schools observed in 1895. "Text-books are valuable in their place," he reported, "but a laboratory system of instruction should be introduced into all studies, notably the sciences, and pupils should handle the apparatus and perform the experiments for themselves."[38]

At the end of the two-year district-wide retrofitting project, the superintendent, Albert Lane, stated proudly that the Chicago "course of study is comprehensive enough to enable a student to prepare for admission to any college or scientific school." And the newly updated science facilities were no small part of this. "All high schools are now equipped with good laboratories for work in biology, physics, and chemistry," he wrote. "They furnish advantages for elementary scientific study superior to those furnished by most of the colleges of the land twenty-five years ago." One teacher at a local high school gushed that the laboratory equipment being provided "causeth the heart of the physics worker to rejoice." "Compared with the little corner room with the surprising induction coil, which some of us enjoyed during our high school days," he noted to colleagues, "such provisions seem marvelous." Chicago, of course, was not alone in making such claims. Sentiments like these were common in school reports and school dedication ceremonies across the country during these years.[39]

By the turn of the century in America, the laboratory method had captured the fancy of a broad array of university scientists, school administrators, and community leaders. Edwin Hall's early work at Harvard developing and promoting his descriptive list of experiments with the help of the standards set out by the CEEB resulted in the Harvard course (or something

Figure 5. Typical early twentieth-century high school chemistry laboratory. Laboratory of the South High School, Omaha, Nebraska, ca. 1930. From the *Report on Equipment, Apparatus, and Materials for Teaching Science in the Secondary Schools of Massachusetts, The Commonwealth of Massachusetts Bulletin of the Department of Education,* no. 8 (1930), 20. Reproduced from a copy in The Ohio State University Libraries.

essentially similar) being disseminated across the country as the "national physics course," with various apparatus companies lining up to provide low-cost equipment manufactured using the Sabine-Hall designs to schools everywhere. Parts of the laboratory course—and modified apparatus—even made their way to the grammar school level. With the encouragement of Eliot and members of the Cambridge School Board, Hall developed a separate list of experiments that could be used with younger students and offered a course on experimental physics free of charge to the local Cambridge grammar school teachers during the 1892–1893 academic year. Leading educators soon began calling for more laboratory work in all the elementary schools.[40]

Enthusiasm for laboratory instruction, somewhat surprisingly, even began to spill beyond the natural sciences. The history conference of the Committee of Ten extended the argument for this new kind of teaching to its own subject, insisting in its report that the "value of history is increased if it is looked upon in part as a laboratory science." Mathematics teachers began advocating for laboratory instruction in their classes, with some specifying in detail the equipment necessary for a properly furnished mathematical lab—everything from T-squares to carpenter's tapes to spherical blackboards. Pratt Institute librarian Mary W. Plummer, speaking at the Brooklyn Women's Club in 1896, argued for libraries to be seen as laboratories in their own right: "In the study of physics and chemistry the laboratory was the place of battle; why not for history, literature, political economy, etc.?" The chemist Albert Prescott commented on this trend in his presidential address before the American Association for the Advancement of Science. While the "educational laboratory was instituted by chemistry," he noted, "experimental study has been adopted in one subject after another, until, now, the 'laboratory method' is advocated in language and literature, in philosophy and law."[41] By the turn of the century, nearly every school subject seemed to cast itself as a laboratory study of one sort or another.

In an 1898 talk celebrating the fiftieth anniversary of the founding of the AAAS, physicist Frank P. Whitman reflected on the origins of laboratory teaching in the United States. He recounted the older traditions of teaching by lecture and rote memorization, identifying the small steps toward student laboratory work along the way, first in Germany at Giessen, then in the United States at Rensselaer, MIT, Harvard, and Johns Hopkins. One of the most "striking features of the last twenty years," he commented, has been "the spread of science teaching by laboratory methods in the secondary schools." With this came a shift to a more abstract concept of educational value or worth rooted in notions of discipline and morality. "Interest in science," Whitman observed, "has passed far beyond the mere interest in its applications." The trend toward laboratory work and the less technical aspects of science had been, in his view, a success story: "There has been an arousing, an awakening, in educational ideas and methods . . . that is little less than marvelous to him who can look back five and twenty years." Yet he wondered aloud at the same time "whether, after our American fashion, the pendulum may swing too far, and the movement bring with it the difficulties which always attend exaggerations." For Whitman it was "too early yet to say." But looking back, it is clear that those words of caution were harbingers of a backlash to come against the faith in laboratory teaching as an unalloyed pedagogical good.[42]

3 Student Interest and the New Movement

JUST AS THE COLUMBIAN Exposition marked the arrival of the laboratory method in American schools with its display of Harvard's innovative instructional apparatus in 1893, the event also provided a forum for new ideas that would ultimately prove to be that method's undoing. As part of the fair's run, the National Education Association helped organize the International Educational Congress. For twelve days in July people from all over the world came together to discuss the state and progress of education. If things had gone according to plan, it would have been there that President Eliot unveiled the recommendations of the Committee of Ten, with its resounding call for more student laboratory work in the schools. But the science groups were late with their subcommittee drafts, which pushed back the release of the overall report to the end of the year. Rather than providing a showcase for Eliot's educational standards, the congress instead served as a launching point for another set of ideas, those of educational psychology and child study.

On the morning of July 26, with the fair in full swing, Clark University president G. Stanley Hall delivered the opening address for the session on experimental psychology. His topic was "Child Study as a Basis for Psychology and Psychological Teaching." In the early 1880s, Hall had done pioneering work on children's thinking that helped establish the new field of educational psychology, and he returned now to meet the growing demand for such knowledge in the schools.[1] In addition to Hall's talk, there were presentations on "The Imaginations of Children," "Child Study as a

Basis of Pedagogy," and "The New Psychology in Normal Schools," among others.

The three-day session was wildly popular. Nearly all the talks were filled to capacity, and in some cases speakers repeated their presentations for new audiences of teachers and administrators hungry for ideas about how to meet the needs of the children surging into the public schools. One observer reported that "the experimental psychologists were asked to speak before sections representing nearly every grade of school work and were received with hearty appreciation when they said 'Study the individual child, learn how he thinks and acts and feels, and then give him freedom to develop in a natural way.'" "Teachers, physicians, parents, and many other friends of education" gave these presenters their rapt attention.[2] There was a sense of something new and exciting coming into view.

At the time, it seemed of little consequence that Eliot missed his original deadline, giving up the spotlight to Hall and the new educational psychology. The Committee of Ten report, when issued that winter, didn't suffer from lack of attention by any means. But in retrospect, Eliot's missed opportunity that summer foreshadowed the radical changes in science education soon to come. In a little over a decade, laboratory teaching would be seen as an instructional approach out of step with the newer currents of educational thinking and ill suited to the shifting demographics of secondary school enrollments. The formal elegance and power of disciplinary knowledge thought to instill mental acuity and personal virtue—characteristics once deemed essential to the flourishing of the republic—were now seen as downright harmful to the proper development of the nation's youth.[3] All this came to a head just after the turn of the century when Hall launched an attack on the paradigm of laboratory teaching, the experimental physics requirements of Harvard and Edwin Hall. The harsh critique leveled at the Harvard course spurred a reexamination of science teaching across all subjects and sparked a new way of thinking about what science really was and what students might learn about its process.

●　　●　　●

The changing fortunes of laboratory instruction can be seen in the microcosm of the Eastern Association of Physics Teachers (EAPT), a professional organization formed at the height of the public fascination with the laboratory method. Inspired perhaps by the strength of conviction in student experimentation that ran throughout the Committee of Ten report, seven Boston-area high school physics teachers circulated a letter in early 1895 seeking other "*bona fide enthusiastic* specialists" who would be interested in having regular discussions of "the best books, methods and apparatus to be used in the teaching of Physics in the Secondary Schools." They promised all comers "absolute freedom of debate" and a program of "short practical papers free from rhetorical flourish." The only hard requirement for membership eligibility, according to the organization's constitution, was that members be "actively engaged in the instruction of Physics by the laboratory method." The group's dinner meetings, held every other month during the school year in local high schools or frequently at the United States Hotel in Boston, held true to the members' commitment to frank discussions—discussions that began to reveal some disenchantment with the prevailing trend of the time.[4]

The early meetings naturally reflected the spirit of the group's founding mission, which was to promote student laboratory work. Edwin Hall made the short trek north from Cambridge to the high school in Malden to speak at the association's November 1897 gathering, lecturing to a receptive audience on the value of the laboratory in enabling students to retrace the intellectual steps of the great natural philosophers who established the foundations of the discipline. At this same meeting, the executive committee of the association reported on the progress of their first publication, a manual of laboratory plans to help teachers and school district officials make the transition to lab-based instruction. The manual, initiated in response to the "frequent inquiries . . . for plans and specifications of the laboratories for Physics which have been fitted up during the past ten years," was issued in 1898 and included descriptions of both ideal and actual arrangements of instructional facilities, apparatus, and supplies, with detailed cost breakdowns of each. Such information was in high demand, the authors noted, as a result of "the policy of Harvard College in placing experimental physics among its requirements for admission." The group was pleased to report in a follow-up survey on physics teaching methods in eastern high schools that 51 percent offered the Harvard preparatory course and nearly 65 percent of all instruction in the area was focused on the laboratory method.[5]

But despite the group's explicit commitment to laboratory work, discordant views began to surface. At the sixth meeting, Cambridge physics teacher Charles Warner laid out his concerns over a too-strict adherence to the inductive method. The success of the Harvard pamphlet was unquestioned. "Backed by the strong arm of the University," he acknowledged, "it was probably the best thing that could be done to enable the physical laboratory to gain a footing in our high schools." He worried, though, that too much emphasis on the method of physics was turning students away. "Witness your experience with the stretching, bending and twisting exercises of the 'forty-experiment' course," he lectured his colleagues. "Is there anything interesting, not to say inspiring about them?" In the group's avowed spirit of frank discussion, he implored the group: "Let us not allow our zealous devotion to the scientific method to kill any natural desire of our pupils to know something worth knowing." Warner wasn't alone. A Charlestown teacher described a "change of heart" in his after-dinner comments to the association in 1898. "After eleven years' personal observation," he deemed the physics teaching in Boston schools a failure. He was convinced that "physics could not be an *experimental* science" for his pupils. Another member concurred, insisting that the physics textbooks of the day, written by "college mathematical physicists," were beyond the abilities and interests of the typical high schooler.[6]

These early countervailing voices signaled a subtle shift in the group's focus. Discussions at association meetings began to include topics other than the ins and outs of laboratory teaching. Some members offered their views on the role of lecture as a means to "broaden the pupil's horizon and to bring him into touch with the great achievements of inventors of all time." Others, feeling that the laboratory method had been "carried in some cases perhaps too far," sought to rehabilitate the "quiz or recitation method" of teaching. In 1900 the association offered a report titled *Methods of Instruction in Physics in Secondary Schools,* its second major publication. An initial survey of area teachers found that the campaign for laboratory teaching had succeeded perhaps too well. "By far the greater number of the teachers of physics returning the circular [gave] prominence and emphasis to the 'Laboratory Method,'" the committee reported, "some believing that it should be dominant and determine the character of all instruction in physics, the other methods being entirely subordinate when used at all." In the end, the EAPT committee offered a more balanced perspective, insisting that "the Lecture and Demonstration, the Recitation, and the Laboratory Method must all be employed in order to secure the best results. . . . No one of these methods should be dominant."[7]

The change in tone from the founding of the association in 1895 to its 1900 report on instructional methods signaled a new willingness to question the orthodoxy that emanated from Harvard and the Committee of Ten. If that willingness came with a measure of caution, at least for some members, about taking on what had clearly become the establishment pedagogical position, that caution was largely stripped away in the spring of 1902 when G. Stanley Hall gave the opening address at the association's thirty-third meeting at the Massachusetts Institute of Technology. In a talk on the nominal topic of manual training and physics, Hall launched into an extended attack on the Harvard physics approach. He led off his address by noting the centrality of physics to all science, its astounding growth, the utilitarian promise it offered civilization, and its unmatched disciplinary value to those who study it. But despite the praises one might sing of physics as a school subject, he noted, it "is now well along in the stages of educational decadence." Hall's appraisal was based on reports from the U.S. commissioner of education that showed high school physics enrollments in steady decline, from just over 25 percent when the Committee of Ten report was published to approximately 19 percent nine years later. In Hall's words, something was "very wrong."[8]

For Hall, the cause could be found in the "neglect and violence done to the nature and needs of the youthful soul." Not only was physics being taught as an impersonal disciplinary pursuit with no recognition of its history or the great heroes who made it, but worse, the "half-score of textbooks . . . I have glanced over," he commented, "seem essentially quantitative, require great exactness, and are largely devoted to precise measurements"—all hallmarks of the Harvard laboratory approach. "Boys of this age," Hall asserted, "want more dynamics. Like Maxwell when a youth they are chiefly interested in the go of things." "The normal boy in the teens," Hall went on, "is essentially in the popular science age. He wants and needs great wholes, facts in profusion, but few formulae."[9] The true benefits science offered, at least as Hall saw it, came from knowledge of and engagement in the new industrial world where students lived, not from some fusty, outdated idea of mental refinement or moral uplift.

While Hall's presence before the relatively small gathering of the Eastern Association might have been a novel event, his criticisms were no doubt familiar to those assembled. Hall's attack on the sterile laboratory experience closely followed the text of a widely publicized address he had delivered seven months earlier at the fall meeting of the New England Association of Colleges and Preparatory Schools. There, in the main hall of the Boston Latin School, Hall attacked the excesses of precision and quantitative experimentation that characterized the national physics course. Drawing on his work in the field of child study, he argued that there was a mismatch between the formalisms of rigorous inductive methods and the natural inclinations of high school youth. "There are two standpoints from which everything can be regarded," he insisted, "the logical and the genetic. One is the method of system, and the other that of evolution." The latter view elevated the interests of children above the logical arrangement and processes of a given discipline. In high school physics, according to Hall, just the opposite was taking place. In current textbooks and laboratory manuals, "the topics are admirably chosen" and "are such models of condensation and enrichment," he observed, that it seemed almost a "perversion that our youth pass it by." From the genetic perspective, however, this made perfect sense. For, as Hall noted, "exactness comes relatively late in the development of the youthful mind."[10]

The point of Hall's critique was to highlight the chasm that appeared to have opened up between the disciplinary structure of academic physics and the everyday world students routinely encounter or even seek out as a result of their natural interests. As much as physicists and physics teachers of the day marveled at the logical coherence and beauty of their field, it existed in isolation from the immediate experiences of the child. For students, there was no clear and obvious path from their everyday world to the heights of disciplinary physics, and Hall believed that a forced march through the process of formal scientific reasoning, as seemed to be the prevailing practice, did far more harm than good. High school students simply were not developmentally ready to engage in those rigors.

•   •   •

With Hall's leadership, the new psychology had taken firm root in the freshly tilled intellectual soil of late nineteenth-century America. Shaped by the prominence of the evolutionary thinking of the time, it encouraged scholars to see the human mind as a legitimate object of scientific study—continuous with the natural world—rather than as something open only to introspective analysis and often tinged with strong moral or religious overtones. While the emergence of psychology as a field of academic study was perhaps a foregone conclusion given the intellectual currents in that era, its development—at least the subfield related to children's minds—received a considerable boost from the rapid expansion of school enrollments that accompanied the urbanization of the United States as well as the crusade against childhood labor around the turn of the century. The many pedagogical and social issues that arose from the large numbers of students, often children of recent immigrants, pouring into classrooms across the nation proved to be fertile ground indeed for the application of theories and methods from the new field. It was in this context that Hall's attack on physics teaching captured the attention of science educators.

Practitioners of the new psychology sought to understand the human mind through the application of laboratory methods of experimentation rather than through individual subjective analysis, as had been the tradition in British moral philosophy. They used novel apparatus to test subject reaction times, the acuity of sensory perception, physiological changes related to mental function, the limits of memory, and the like. It was a hardscience approach introduced to America through the efforts of Harvard's William James, who was inspired by the German research ideal and specifically the experimental work of Wilhelm Wundt at the University of Leipzig. James fashioned an institutional presence for the new discipline at Harvard in the 1870s and, after twelve long years of writing, published his monumental, two-volume *Principles of Psychology* in 1890. The book, which sought to place psychology on the same level as the natural sciences, laid down the intellectual foundation of the field in the United States.[11]

While James introduced the new psychology to the American academic community, G. Stanley Hall's efforts helped cement the field's professional identity. After earning his PhD under James at Harvard in 1878 (the first psychology PhD in the United States), Hall traveled to Europe for additional training and returned to the United States in 1880, taking a Harvard lectureship in philosophy and pedagogy at Eliot's invitation in 1881. His pedagogical lectures there drew large audiences of Boston-area teachers. Determined to make a career as a professional psychologist rather than a pedagogue, he moved to Johns Hopkins the following year, where he established the first American laboratory of experimental psychology. In the coming years, Hall had a hand in the creation of nearly all the professional apparatus of the new discipline. He started the *American Journal of Psychology* in 1887 and, beginning a move back toward the study of teaching after assuming the presidency of the newly established Clark University in Worcester, Massachusetts, founded the journal *Pedagogical Seminary* in 1891. The following year Hall organized the American Psychological Association, which he led as its first president.[12]

When he appeared before the International Educational Congress at the World's Fair in Chicago two years later, he came fully credentialed as the preeminent authority on the subject of children and learning in the United States. And he addressed an audience primed to hear his message. The large crowds Hall drew to his lectures at Harvard in 1881 and the educational courses Eliot organized in 1890 (from which sprang James's popular *Talks to Teachers on Psychology*) testified to the demand for a science to inform education, if one could be had. In Chicago, the overflowing crowds were yearning for an understanding of the contents of children's minds and the methods of child study. The seemingly unconditional adoption of the central ideas of this new field by the professional education community led some psychologists themselves to caution against the unchecked optimism that seemed to be running rampant. "A warning ought to be sounded to the teachers against their rush toward experimental psychology," wrote Harvard psychologist Hugo Munsterberg in the *Atlantic Monthly* in 1898. "This movement began as a scientific fashion. It grew into an educational sport, and," he scolded, "it is now near the point of becoming an educational danger."[13] Such warnings did little to dissuade enthusiasts, particularly those who were captivated by Hall.

Over the course of his colorful career, this was Hall's greatest success: selling the potential of experimental psychology in education, particularly through research aimed at understanding the psychological development of children. Here was an approach, grounded in the reigning excitement of evolutionary biology and experimental science, that could provide real insight into the process of education and human progress. "The newest and perhaps richest field for psychology," Hall wrote in an essay for *Harper's Monthly*, "is now opening in the studies of evolution, which are just beginning to be applied to the human soul." He went on to conclude that child study, as one example of such application, "is very slowly but surely giving us a new and more solid basis for education."[14]

Hall's choice of school science as the target of his attack, first voiced at the meetings of the EAPT and the New England Association of Colleges and Preparatory Schools, offered a perfect nexus of conflicting social and intellectual trends that served to illustrate the failure of a pedagogy grounded in a rigid view of abstract science and its inductive method. The most obvious of these to Hall and teachers across the country was the rapidly expanding number of children in the schools. In 1886, the year Edwin Hall assembled his descriptive list of exercises at Harvard, there were fewer than 4,000 high schools in the United States, both public and private, accommodating just over 160,000 students. Over the next decade and a half, high school enrollments skyrocketed, increasing to more than 655,000 by 1901, the year G. Stanley Hall addressed the New England Association. Over the span of only fifteen years, student numbers had quadrupled at a time when the general population increased only by little more than a third. In 1903 the commissioner of education talked about the "high school movement" and illustrated it with graphs that vividly portrayed this rapid expansion. During this period new high schools (none complete without state-of-the-art scientific laboratories) were built on average at the rate of one every day to meet the demand for space. By 1910, the total number of new schools stood at nearly 12,000.[15]

Given the ever-expanding student population and the increasing prominence of science in the intellectual and industrial world around the turn of the century, it was not unreasonable to expect a subject such as physics to grow in popularity. The fact that physics was losing ground to other school subjects, however, as Hall repeatedly pointed out, suggested that there was something wrong with the way it was being taught, that there was a mismatch between the subject and students. The demographics of the high school student body had changed dramatically. Children of all classes, from farmers to tradespeople to unskilled laborers, filled the desks and laboratories of the new schools. Although it had always been true that high schools, even in their early days, were only marginally concerned with preparing students for college, from the 1880s through the end of the century the percentage of students preparing for college declined significantly. Commenting on this downward trend, one observer (clinging to the inaccurate college-preparatory view of high school) stated that the numbers "show at a glance that the problem of the secondary school has changed from that of the fitting school to one of a decidedly *unfitting* school; a school in which only 6.8 per cent of the pupils anticipate college work of any sort." Looking back on that period, Charles Hubbard Judd, the director of the University of Chicago School of Education, summed it up best: "We have brought into the secondary schools of the country a great body of new people, people with entirely

DIAGRAM 1.—*Number of secondary students in public and private secondary schools.*

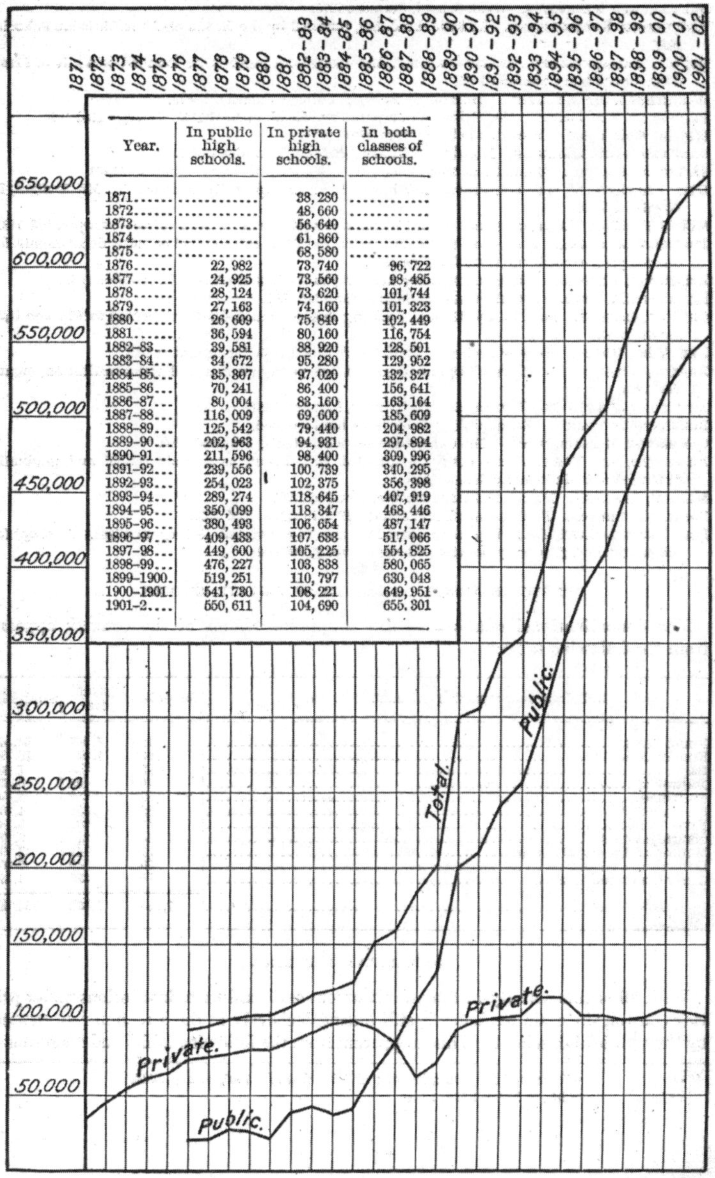

Year.	In public high schools.	In private high schools.	In both classes of schools.
1871......	....	38,280	....
1872......	....	48,660	....
1873......	....	56,640	....
1874......	....	61,860	....
1875......	....	68,580	....
1876......	22,982	73,740	96,722
1877......	24,925	73,560	98,485
1878......	28,124	73,620	101,744
1879......	27,163	74,160	101,323
1880......	26,609	75,840	102,449
1881......	36,594	80,160	116,754
1882–83...	39,581	88,920	128,501
1883–84...	34,672	95,280	129,952
1884–85...	35,307	97,020	132,327
1885–86...	70,241	86,400	156,641
1886–87...	80,004	83,160	163,164
1887–88...	116,009	69,600	185,609
1888–89...	125,542	79,440	204,982
1889–90...	202,963	94,931	297,894
1890–91...	211,596	98,400	309,996
1891–92...	239,556	100,739	340,295
1892–93...	254,023	102,375	356,398
1893–94...	289,274	118,645	407,919
1894–95...	350,099	118,347	468,446
1895–96...	380,493	106,654	487,147
1896–97...	409,433	107,633	517,066
1897–98...	449,600	105,225	554,825
1898–99...	476,227	103,838	580,065
1899–1900.	519,251	110,797	630,048
1900–1901.	541,780	108,221	649,951
1901–2....	550,611	104,690	655,301

Figure 6. Graph of the number of secondary students in public and private secondary schools from *Annual Reports of the Department of the Interior for the Fiscal Year Ended June 30, 1903; Report of the Commissioner of Education* (Washington, DC: Government Printing Office, 1904), 1:566. Reproduced from a copy in the author's collection.

GROUP OF A PORTION OF THE PEOPLE WHO ATTENDED THE ST. LOUIS MEETING OF THE CENTRAL ASSOCIATION OF
SCIENCE AND MATHEMATICS TEACHERS.

Picture taken from main entrance of McKinley High School.

Figure 7. Attendees of the 1907 Central Association of Science and Mathematics Teachers (CASMT) meeting held at McKinley High School in St. Louis, Missouri. From Florian Cajori, "Lessons Drawn from the History of Science," *School Science and Mathematics* 8, no. 2 (1908): 85–97, photograph on page 97. Reproduced from a copy in the University of Wisconsin Libraries.

different motives for attendance upon those high schools . . . from the motive that prompted people going to the secondary schools twenty-five years ago."[16] Quantitative physics just wasn't a draw for these students.

The mismatch between the everyday needs of students and the formal presentation of academic disciplines was all too evident to Hall. In 1904 he persisted in his attack on high school physics teaching in his landmark, two-volume work *Adolescence.* The laboratory emphasis on precision measurement and quantitative analysis that had come to characterize the inductive method of science and learning in the universities was simply too much for the masses of students that Hall referred to as the "great army of incapables." Teachers, he argued, needed to understand that students of this age yearn for stories about the great heroes of science and prefer the action of mechanical devices over mathematical formalisms. "Last, and perhaps most important of all," he explained, "the high school boy is in the stage of beginning to be a utilitarian."[17] With this, Hall resurrected the debate over the practical versus the disciplinary value of science teaching that advocates of pure science thought they had put to rest with the triumph of the laboratory movement.

•   •   •

The growing dissatisfaction with laboratory instruction in physics wasn't limited to the East Coast. A second and ultimately more powerful center of opposition emerged in the Midwest under the leadership of Charles Riborg Mann, a University of Chicago physicist who, like his counterparts in the Boston area, drew on the ideas of G. Stanley Hall and the new psychology to argue against the overemphasis on laboratory teaching. Mann's efforts sparked what came to be widely known as the New Movement in Physics Teaching. The New Movement, launched in December 1905, was bolstered by and contributed to the growth of the Central Association of Science and Mathematics Teachers (CASMT). Established only two years earlier, CASMT emerged as the leading force for the professionalization of science teaching in this period. This new group of high school science educators largely cast its lot with the increasingly popular psychological theories of the time that viewed student interest as the key to meaningful learning.[18]

The New Movement got its start in response to the July 1905 publication of an NEA-recommended, updated list of experiments for high school physics, a list not very different from those found in the Harvard pamphlet.

Members of CASMT mulled whether to endorse that list at their December meeting in Chicago and elected instead to appoint a committee to look into the matter more closely. Mann took charge and was joined by two physics teachers, Charles Smith, from Hyde Park High School in Chicago, and Charles Adams, from Detroit. The group decided to canvass teachers across the country via a series of circulars that asked about the best means of teaching high school physics. The first circular included an extended list of nearly a hundred typical laboratory experiments (far more than contained on the NEA list) and requested that respondents indicate which of these were essential for high school students and why. In subsequent circulars, the group expanded the scope of its survey to ask about the amount of time that should be devoted to laboratory work, what year in high school physics should be taught, whether definitions should be stated up front or derived from concrete examples, and so on. In looking across all the questions posed, it seems clear that the committee's intent was to cast the existing curriculum in the harshest light possible and invite readers to suggest any and all alternatives to the status quo.[19]

As the teachers' responses came in, the group began summarizing the results. Although Mann conceded that the replies indicated that "physics teachers are far from agreed as to the aims, methods, and needs of their work," this didn't stop him from advancing a set of guidelines that he hoped would jolt high school physics teaching out of the traditionalist rut he believed it was in. These included calling for a reduction in the number of topics addressed in the year-long course (eliminating specifically "the more abstract, and technical topics . . . that have no bearing on the student's life"), a reduced emphasis on precision measurement, and the inclusion of more qualitative laboratory activities tied directly to everyday experience.[20]

However ambivalent the collective opinion of physics teachers in the schools may have been, Mann's views on the state of high school physics were crystal clear. Invoking the ideas of G. Stanley Hall, he repeatedly denounced the Harvard model of laboratory teaching and with it the recommendations of the Committee of Ten. "We have fallen heir to a set of arid, parched, and lifeless experiments, and to a stock of laboratory apparatus for performing the same," he insisted. "The whole thing resembles more a mummy than a living man, and its only just place is in a museum." A sympathetic colleague, perhaps with an eye on all the new schools going up, complained about the "lavish waste of money in the purchase of high-priced and utterly unsuitable apparatus." Mann believed as well that far too much effort had gone into getting the equipment, procedures, and notebook requirements just right without giving due consideration to the students. In a talk before the New York Physics Club in June 1906, he complained that "it seems at times as if we have been so busy perfecting our apparatus and methods of teaching physics, that we have for the time being entirely lost the art of teaching boys and girls." "We seem to forget," he added, "that youngsters of the high school age are interested primarily in life, in growth, in activities, and, as President Hall puts it, in the 'go' of things."[21]

Mann's reference to G. Stanley Hall in this address was typical of his activism during this period. Although he did not arrive in Chicago until three years after Hall's command performance at the Columbian Exposition, Mann was nonetheless swept up in the psychological ideas then gaining currency. He pointed out that the Committee of Ten had stated plainly that the primary function of the secondary schools was to prepare boys and girls for the duties of life. But, he asked pointedly, had they? "Have there yet been developed methods of instruction which, in the light of modern psychological investigation, are known to be best suited to meet present life needs of pupils? Has experimental psychology been invoked to prove that the methods in use are the best ones attainable?" In his talks and papers, he encouraged his audience to immerse itself in the work of Hall and the new psychologists and to see whether current teaching practices were aligned with what researchers had come to understand about student learning. By 1912, he was firmly committed to the new psychological thinking. In a book on the teaching of physics, he denounced the outdated faculty psychology and mental discipline of the prior era. "A new day is dawning for the school children," he exclaimed. "The science of psychology is coming to their rescue, by proving that the human mind is not made up of separate faculties of which reason and memory are the chiefs."[22]

Together, science educators such as Mann, with the support of their new professional associations, made the overthrow of traditional, laboratory-based physics teaching a national cause. In the sixth and last circular published under the auspices of the Central Association (now along with support from the North Central Association of Colleges and Secondary Schools and the New York State Science Teachers' Association), Mann announced the organization of a symposium to address the question "What should be the purpose of the instruction of physics in the secondary schools?" Three

categories of professionals were invited to respond: psychologists and educators, college physicists, and secondary school men. The list was a who's
who of eminent figures from each field. Among the first group were university presidents Nicholas Murray Butler from Columbia and G. Stanley Hall
from Clark as well as Harvard's William James and John Dewey from the
University of Chicago (both as much philosophers as psychologists). The
physicists included Chicago's Nobel laureate Albert Michelson and Robert
Millikan, who would win his own Nobel Prize fifteen years later. The
schoolmen consisted of the key leaders of the New Movement such as Mann
and John Woodhull, a science educator from Teachers College, Columbia
University.[23]

Although their recommendations were not fully aligned, the consensus
(a foregone conclusion, really, based on the invitation list) was that the highly
quantitative and abstract laboratory approach to teaching had run its course.
A new focus on student needs and interests was necessary to revitalize high
school physics education. "Far too much has been made in recent years of
accuracy of measurement in the teaching of elementary physics," Butler wrote
in his contribution to the symposium. "It is much more important to throw
emphasis upon the descriptive aspects of the science and to feed the growing
mind with food that really interests it and helps it grow, than to pursue the
will-o'-the-wisp of training in some imaginary power of habitual accuracy."
Woodhull reiterated the sentiments of other reformers, insisting that "nothing
has so retarded the progress of physics teaching as the idea that it is essentially a science of measurements and that its chief function is to train pupils
in accuracy by means of exact measurements." The "dusky villain" in all this,
an early historian of the struggle concluded, "was our old friend, the disciplinary theory of education."[24]

Despite the vigor of the campaign, the movement against laboratory instruction didn't proceed entirely without objection. There remained a handful
of physics educators who defended the popular lab-based pedagogy of the
late 1800s. They clung to the older views of student learning and argued further that the central moral component of such work was in danger of being
lost. In the opinion of one stalwart, "Every physical laboratory should be a
spot which the future man may look back upon as one of the places where
he learned that the really serious matter in life is to meet each individual
responsibility as it comes along." The moral benefit came, traditionalists believed, from the laboratory exercises being the closest thing to active scien-

tific thinking itself: "The success of science teaching rests on its ability to train minds, to develop mental power, along just those lines which we have characterized as scientific thinking—and to do this is the spirit of the laboratory method." "To do away with laboratory instruction," a critic insisted, would result in a "total misconception of the nature of physical inquiry and of the actual method of scientific procedure."[25]

Others derided the new focus on student interest by comparing it to the kindergarten movement, which was gaining popularity around the same time, particularly in and around Chicago. In this new "kindergarten spirit," one physics teacher complained, "boys and girls must be amused, must be coddled and given a good and easy time in school or the up-to-date parent will know the reason why, and teachers will be arraigned . . . for starving their poor minds, torturing them perhaps before their time, giving them bread when their cry was for candy." An East Coast college professor made this point as well. "Our schools are not kindergartens," he stated. "Science teaching has a definite and difficult problem before it and its mission is not furthered by listening to every educational fad that clouds the horizon." Others piled on, and "kindergarten method" became a frequently invoked pejorative phrase of the day.[26]

Edwin Hall, the widely acknowledged architect of the Harvard approach, watched with interest as the New Movement unfolded. He wasn't completely unsympathetic to the issues being raised. Ironically, Hall himself never argued for the laboratory course on mental-discipline grounds, or at least claimed not to in retrospect. "As to the notion that physics should be studied mainly for mental discipline rather than for its own value," he quipped, "I cannot help comparing it to the notion of a man, well known to my college days, who played the violin for the purpose of strengthening his arm. The tune was of little consequence provided it gave him plenty of exercise." Any such "salutary effects" (whether muscle development or mental discipline) should be secondary, he agreed. Moreover, Hall insisted that the principle that guided his selection of laboratory exercises for the original list had always been that of "practical utility" rather than some adherence to an abstract notion of mental training.[27]

He would go only so far, though, to soft-pedal his earlier views. When push came to shove, he balked at the reformers' insistence on reframing the curriculum in light of student interests. "We cannot properly take as our guide in teaching the unintelligent wishes of these pupils," he stated plainly.

"We must not adopt or encourage the Sunday newspaper attitude toward physical science," referring to the comics and games newspapers published for the entertainment of children on the weekend. Nonetheless, despite his personal commitment to the Harvard model, he seemed to recognize that the approach had its limitations. "Although it has done much good for the teaching of Physics throughout the country during the past ten or twelve years," he wrote in an internal Harvard committee report, it "is now generally admitted to be too exclusively a laboratory course."[28] With that assessment most everyone agreed.

Although the EAPT- and CASMT-based New Movement led the highly publicized charge against the laboratory method in physics, leaders in the other subjects were similarly eager to leave the laboratory behind in the nineteenth century, where it had originated. An instructor at Harvard, of all places, claimed that chemistry too "suffered from the irrepressible wave of laboratory madness which has swept over the whole educational world. Laboratory work has been carried far beyond its limits and things have been expected of it which it never did and never can do." Charles Bessey, the early champion of such teaching in botany, felt that the approach had been overdone in the life sciences as well. Its pervasive adoption in schools was, he commented, "a serious error of judgment, and had led to no end of wasteful educational experiments." "These days we hear a great deal about the Laboratory Method," mused another educator. "Sometimes I think that the teaching world has gone daft on the subject, for have they not tried to apply this method to the teaching of every known subject from cube root to Caesar?"[29]

· · ·

American education in the early 1900s could be described, perhaps charitably, as existing in a state of ambiguity and instability. Indeed, "storm" and "fermentation" metaphors were often invoked to characterize the assorted movements and discussions taking place concerning the school curriculum. These extended to the broader role of education in society at the time as well.[30] In science education, the competing forces of scientists and college faculty on one side and a newly professionalized group of educators on the other clashed on a rapidly changing field of practice. Institutionally, formal schooling was expanding as never before, especially at the high school level, which resulted in dramatic shifts in student demographics as more—and different kinds of—students ambled in to fill the new seats, as we've seen. This broader cross section of students, reformers argued, were seeking something other than traditional academic preparation from their classroom lessons. And with this new clientele in mind, those in the vanguard began turning away from the entrenched laboratory teaching approach in all the major school science subjects.

The confidence with which the science educators took leave of the laboratory can be attributed to a number of factors. The new psychological theories about the importance of student interest that G. Stanley Hall brought to public attention were, of course, key to the shifting ideas about teaching. Advocates of science instruction in the general school curriculum once had insisted that it provided the same level of abstract mental training as Latin, Greek, and mathematics; now a new generation of advocates argued against the idea of intellectual development of this sort. The faculty psychology of the 1800s gave way to the new psychology of the twentieth century, so much so that by 1906, the Columbia University educational psychologist Edward Thorndike could say with conviction that "any method of teaching which has no other raison d'être than its supposed disciplinary effect is almost surely questionable."[31] Theories of student interest and new developmental constructs such as "adolescence" provided firm intellectual ground on which the science education reformers could build their new educational approaches.

The professional educational network that took shape during this period was another factor that enabled the move away from the laboratory. The success of the reform efforts—their emergence as a "movement," in fact—was tied directly to the growing number of associations like the EAPT and CASMT. The forums provided by the regular meetings of these various groups and the wider audiences they could reach through new professional journals enabled reformers to coordinate their efforts and enlist supporters across the country. Mann, for example, was far from the only voice in the New Movement chorus for reform. Other leading educators took up the call as well. Prominent individuals who stood alongside Mann included H. L. Terry, the Wisconsin state high school inspector; George Ransom Twiss, a Cleveland physics teacher and textbook author; and John Woodhull, from Teachers College. A teacher from Stuyvesant High School in New York characterized the movement as a "perfect cyclone, with its storm center not far from Chicago University," and acknowledged that much of its power came from the "many scientific and educational societies scattered from San Francisco to New York [that] have lent their names to [its] promotion." Mann recognized this too. "By uniting the teachers for the solution of their common problems," he noted, these associations "have been a source of inspiration to all of their members."[32]

The rise of a professional education establishment in the years around the turn of the century coincided with the dramatic expansion of public education. The growth of high school enrollments and the infrastructure that grew along with them (including not only schools but also textbooks, interested parents, teacher education programs, and university schools of education) provided a base of social and political support on which educators could rely to advance their professional interests. "Within recent years the public high schools have become the most important educational institutions in the country. They surpass the colleges in buildings, laboratory equipment and teaching force," wrote one reformer in 1907. Questions about why "high school teachers have no professional status" gave way to increased recognition that this new group of educators could bring about real change. The security their professional network afforded no doubt emboldened them to distance themselves from the old-line university science educators. And in doing so, they came to accept the fact that their role, primarily as teachers, was fundamentally different from that of their university-based counterparts. They had come to recognize, as one put it, "the true mission of the high school."[33]

Most significant in all this was the changing characteristics of the students who began coming to school in the new century. The fact that they were seen as needing something other than strictly academic instruction was clearly a driving force for change. The new generation of science educators viewed continued efforts to impose an abstract, research-based laboratory science onto the masses as wrongheaded and even injurious. "The teacher who tries to introduce into the high school a miniature of the course which he has been through in college, what a misfit he is!" exclaimed one observer. Even soon-to-be Nobel laureate Robert Millikan felt that the "creeping over of the methods and instruments of research and specialization from the university into the high school"—where they "have absolutely no place," he added—was the "gravest danger" for science education.[34]

But as much as this point in history was defined by educators steering the science curriculum away from abstraction and formal laboratory methods, it was in many ways equally defined by the scientific community doing just the opposite. As university science faculty embraced the German research ideal in places such as Johns Hopkins, Clark, Harvard, Chicago, and Stanford, they pushed forward an agenda of narrow, specialized scientific study. In the closing decades of the nineteenth century, an array of new scientific associations (such as Hall's American Psychological Association) sprouted up in the country's intellectual centers to promote work in new disciplinary fields. This fragmentation of scientific work was evident in the organization of the AAAS, the subject sections of which multiplied exponentially from its founding in 1848. Institutional and organizational arrangements such as these reflected a remarkable level of increasingly narrow research activity. The rapid growth of specialization pervaded nearly every aspect of higher education and the research enterprise in this period.[35]

Worries about overspecialization were voiced by education reformers, who saw it as a challenge to teachers and students alike. A common refrain was that, as one small-college president put it, "science in our high schools has become altogether too differentiated, too technical, and too destitute of real attractiveness for secondary-school pupils." On the other side of the desk, the "evil of early specialization," Woodhull insisted, not only "failed to qualify young men for teaching, but there has grown up along with it a distaste for and even a disrespect for teaching." Some went so far as to suggest that the future of science itself was at stake, that the intellectual distance between the researcher and the public had grown to an untenable level. "The work of the specialist," warned a California high school principal and science educator, "is to-day endangered by a lack of general scientific knowledge on the part of the public it aims to serve."[36] So even as the education reformers of this period worked to shift the science curriculum toward the everyday needs and interests of students, the growing American scientific research community was moving ever deeper into itself and away from what the public generally knew and understood.

Science teachers and educators, particularly those toiling away in classrooms, were essentially caught between two worlds—mass education on one side and research science on the other—and it was all they could do to manage. As Mann recognized in 1905 just as the New Movement was getting under way, "The teacher of Natural Science seems to occupy a particularly difficult position." He not only had to strive to realize the goal of successfully engaging all of the pupils under his charge but also had "to deal with a subject that is . . . growing on the material side at an extraordinarily rapid rate. Hardly a month passes in which some new discovery is not announced or some new theory brought forward. How can he ever keep track of it all, let along introduce it effectively into his instruction?"[37]

The greater challenge for teachers, however, was figuring out which method of instruction to use to reach their students. At a time when scientists and other public figures—people such as Simon Newcomb, William James, and the British mathematician Karl Pearson—were calling for a greater embrace of the scientific method for both intellectual and social progress, many classroom teachers struggled with what to do. For more than two decades beginning in the middle 1880s, educators had accepted as axiomatic that the methods of the inductive sciences were the preferred means by which the student should learn, and that "herein lies the justification for the large developments of the laboratory idea." The close identification of these ideas was widely accepted. "Whatever the phrase may be understood to mean," wrote a Chicago astronomer, "the 'scientific method' connotes at least two important notions: . . . induction as a thought-process and laboratory instruction as an external agency in founding and facilitating this process."[38]

Science education had reached a crisis point. With the laboratory method found wanting in light of the new psychological ideas about student learning, the ability of teachers to promote the scientific method and attend to student interest at the same time was increasingly in doubt. Such conflicting feelings were evident in survey comments published in 1910 by George Hunter, an educator and textbook author from New York. Among the responses to a question about whether a high school biology course should place more emphasis on "training in science method" or on the "utility value of the science"—those things that would apply to daily life—one teacher from Illinois wrote: "By all means give the 'utility value' first place. All modern instruction tends toward 'science method'; in fact laboratory methods have become so prevalent in many departments that pupils are found to be deficient in concrete knowledge." Another commented that "utility value is certainly very important," but added, "I confess that I see no reason why the science method should not be the backbone of such a utility course. The problem is how to combine the two." He went on, "I am sure that I have sacrificed too much to science method." Another teacher, from Cincinnati, indicated that, at least for him, the tide had turned: "If by 'science method' laboratory technique is meant, then the utility value seems more important."[39]

Given the choice between "scientific method" and student interest (the very framing of the question presupposed that the two were incompatible), educators increasingly elected the latter, and the form of instruction shifted accordingly. Mann, as part of his ongoing campaign against the Harvard course of study, argued that science teaching for general students should involve "no laboratory work, but plenty of demonstration experiments, lantern slides, photographs, and collateral reading."[40] These were the kinds of techniques that could best interest students and communicate practical information that would be useful in their lives.

Teachers across the country increasingly agreed. At the December 1909 meeting of the New York State Science Teachers' Association, Syracuse University biologist William Smallwood reported on the latest tendency, "the recoil from extreme laboratory methods." Teachers he had surveyed "from Colorado to Maine" expressed their desire "to limit the amount of laboratory work in secondary science, and to return to the earlier plan of a considerable number of demonstrations." Despite this turning away from laboratory instruction, they remained committed to teaching, in some way, the power of scientific thinking to the next generation of school students. "The coveted results lie within the reach of every one of us," Mann exclaimed. "We need no new laboratories, no new apparatus, no new equipment for the work. All that we need is a new method of presentation."[41] That new method was soon to arrive.

4　The Scientific Method

ON A SNOWY BOSTON afternoon in late December 1909, John Dewey stepped to the podium in the physics lecture hall in the Walker Building on the MIT campus to deliver the vice presidential address to the assembled members of the American Association for the Advancement of Science. In his talk, originally titled "Science as a Method of Thinking and as Information in Education," he attacked head on the perceived incompatibility between teaching scientific method and attending to student interest, the dilemma teachers struggled to reconcile in the first decade of the new century.[1] It was the first substantive address from the new educational leadership at the AAAS and became a touchstone for science education reforms to come.

Dewey's presence at the meeting—and indeed in his role as head of the education section—came by invitation. The disarray and confusion science teachers were experiencing over how best to teach science was felt within the professional science establishment as well during those years. Three years earlier in New York, in fact, AAAS president William Henry Welch had announced the formation of Section L, a new interest group devoted to the educational issues facing the sciences. And rather than choosing a senior scientist to head it, he looked to leaders from education. U.S. commissioner of education Elmer Ellsworth Brown, a teacher by training, was enlisted to launch the section in ceremonial fashion at the 1907 meeting in Chicago, and leading the slate of officers chosen for the following year were John Dewey as vice president and Charles Riborg Mann as secretary—two reformers of the highest order. The selection of Dewey as section vice president was a bold move by the association's leadership, signaling perhaps the anxiety many felt about the future of science education in America. But Dewey was a rising star, well known in the field of education and highly regarded for his efforts to make schooling relevant for students. He seemed the perfect person to change the fortunes of science in the schools, and he did not disappoint.[2]

Dewey began his talk by admitting his lack of standing among those gathered: "One who, like myself, claims no expertness in any branch of natural science can undertake to discuss the teaching of science only at some risk of presumption." But despite the significant gap between "those who are scientific specialists and those who are interested in science on account of its significance in life," he insisted that they could find common cause. What they could rally around, he argued, was making the scientific method the central aim of science teaching. "Students have not flocked to the study of science in the numbers predicted," he granted, "nor has science modified the spirit and purport of all education in a degree commensurate with the claims made for it." The reason for this, in his words, was that "science has been taught too much as an accumulation of ready-made material, not enough as a method of thinking."[3]

Drawing from work in psychology, Dewey then moved to shift the commonly held view of the scientific method. Rather than see it as closely tied to the material techniques of scientific work, he presented the essential nature of science as a mode of thought, a way of reasoning about problems and ideas. In making his argument, he aimed to sever the long-standing connection between the method of thinking and laboratory work. He challenged the widely accepted belief that "the student must have laboratory exercises." "A student may acquire laboratory methods as so much isolated and final stuff, just as he may so acquire material from a text-book," Dewey lectured. "One's mental attitude is not necessarily changed just because he engages in certain physical manipulations and handles certain tools and materials," he went on. "Many a student has acquired dexterity and skill in laboratory methods without its ever occurring to him that they have anything to do with constructing beliefs that are alone worthy of the title of knowledge."[4] What was necessary was for students to see science as a way of reasoning.

Dewey's comments that winter day were no doubt strongly shaped by the ideas in his new book *How We Think,* a slim volume he had just wrapped up work on and was soon to publish. The book, targeted to teachers and other educational professionals as a pedagogical guide of sorts, quickly became a bestseller and fundamentally reshaped public perceptions of the scientific method. Its characterization of the psychology of everyday problem-solving was seized upon by readers as an alternative to a scientific method cast in terms of formal logic and so closely associated with laboratory practice. The ideas Dewey outlined in that book recast the way in which scientific knowledge production was taught in schools from 1910 until the middle of the twentieth century (if not later). His problem-solving method, it turned out, was the "new method of presentation" teachers had been waiting for.[5]

•   •   •

Dewey wasn't the first to suggest that an understanding of the scientific method be the central goal of science teaching. Neither was he the only one to see the troubles that resulted from the entangling of the pedagogical commitment to laboratory work with the intellectual process of knowledge production. Attempts to clarify the essential elements of each had been made by scientists going back to the late 1800s. But the identification of scientific thinking with the concepts of formal logic (concepts that proved unappealing and ill-suited to the young minds of the early twentieth century) hampered nearly all these efforts. One of the more notable scientists commenting on the distinction between the two was the ecologist Stephen Forbes. Forbes worked as a school science teacher in Illinois in the decade after the Civil War and later became curator and director of the Illinois State Laboratory of Natural History. He went on to earn a PhD in zoology at Indiana University in 1884 before taking a faculty position at what would become the University of Illinois. No doubt owing to his personal experience as a teacher, Forbes devoted considerable time throughout his career to supporting school science teachers through the provision of biological materials for instruction, offering summer school courses in natural history, and actively participating in educational conferences and meetings.[6]

Forbes's messages to teachers in the years around the turn of the century often focused on the importance of teaching students the method of science. The value of this, he insisted, extended well beyond science. "In all one's personal activities, in his business and professional occupations, in his more or less speculative reflections, in the general ordering of his mental life, in all that pertains to him, indeed, which contains a rational element," he stated, "the scientific method has its value and its use, for it is simply the method of right reasoning applied to matters of fact."[7] Forbes's reference to "right reasoning" followed the lead of T. H. Huxley, who commonly spoke of the scientific method as nothing more than "organized common sense," and in this way Forbes sought to make a strong argument for seeing scientific method first and foremost as a process of sense-making.

In the opening address before the New York State Science Teachers Association meeting in Rochester in 1900, he offered a definition of this method and laid out his vision for how it might best be taught. He began by making clear that it had nothing to do with the physical aspects of scientific practice. "What shall we mean by method in this discussion?" he asked. "Not the mere use of tools of any sort, however complicated and invaluable; not the manipulation of apparatus, or any form of mechanical operation on anything." He explained that these things may help with making observations, "but they do not in the least help to organize the facts accumulated, or to reason on them when organized." The scientific method was a "mental method." "The subject," he affirmed, "is thus not physical but psychological."[8] He went on to walk the audience through the various stages of reasoning scientists engaged in—the accumulation of facts, their bundling and classification, the development of hypotheses, and the deduction from those hypotheses of specific consequences that must be tested by observation. His characterization of scientific method was not very different from that offered by Karl Pearson in his popular 1892 book *The Grammar of Science,* which presented the essence of the scientific method as the classification and logical ordering of facts.[9]

The process of reasoning Forbes described was light-years from what typically happened in classrooms that purported to teach how science worked. "Take, for example, the laboratory method in biology," he insisted. "The student under instruction by this method sits at his table with a lifeless object before him of more or less complicated structure, and a book beside him which is essentially a manual of directions as the mechanical routine of his work, a nomenclator of the parts of the object under his examination." This, in Forbes's view, hardly exemplified the method of science. The student "reads the book, he does the things he is told to do, he looks at the things he is told to see, he observes and records and draws, and he listens to the remarks of his instructor, and in it all he does not so much as lift one foot from the earth on which rests the lower end of the ladder which we call the scientific method." Activities in the well-supplied physical and chemical laboratories of the day were much the same—the student's "procedure is practically dictated, and his result is predetermined." It was all "largely a drill in mechanical operation rather than a practice in mental method."[10]

As insightful as Forbes's critiques of laboratory work were, his recommendations for how teachers might best teach that process relied on traditional appeals to instruction in formal logic. Speaking before a group of teachers in 1903, he suggested that at the college level "we might perhaps improve our work by correlating our different science departments with . . . the department of logic," noting that "a great part of every serious scientific research is nothing more or less than applied logic." "Any student specializing in science," he explained, "should be expected to study this logic, so taught, so thoroughly as to master it as a practical instrumentality in his scientific work." His recommendation for instructing high school students on this topic was much the same, though he recognized the challenge it posed. "Some of you I fear," he said to the teachers in the audience, "are even now suppressing your smiles as you think of introducing practical logic to the athletic boys and the romantic girls of your high school classes."[11] Yet younger students are no less capable of mastering these ideas, he argued. The teachers' smiles seemed to suggest otherwise.

Others scientists and educators, in one way or another through the years, followed the line of argument Forbes advanced. Calls for training in logic as a foundation for understanding scientific reasoning or the repeated analysis of such thinking into the steps of formal logic were not uncommon. Back in 1876, Simon Newcomb wrote that the country needed "the instruction of our intelligent and influential public in such a discipline as that of Mill's logic, to be illustrated by the methods and results of scientific research." More than a quarter century later, the authors of a pedagogy textbook out of New York lectured would-be teachers that "the fundamental basis of the general method of science is observation of particular facts and drawing conclusions from them. This is induction as a process of logic; and its prominence in science has led to the use of the term 'inductive' method as synonymous with 'scientific' method." They were quick to state, though, that such work involved "deduction and verification" along with "the logical process of induction." Formalisms were the order of the day, and these formalisms were deployed in the service of student mastery of specialized disciplinary knowledge, whether the students liked it or not.[12]

●   ●   ●

Things began to change in the years leading up to Dewey's address in Boston. From the crowd of scientists and science teachers steeped in the ideas of the previous century, Charles Mann, for one, stepped forward with a different vision of what kind—or what essential elements—of scientific method might be taught. Recognizing that the scientific method, "if followed in too much detail, is found to be very complex," Mann insisted that there were nonetheless several things about it that were "of great value to science teaching to understand." The most important of these, he argued, was to see the scientific method as "but a part of a general process of human action"—as a tool "to be used in the service of a strong purpose." For Mann that purpose was something to stir up within the child, not something determined in advance by the instructor, and it certainly wasn't mastery of "one of our readymade systems as described in a science text."[13] Mann's conceptualization of scientific method as a purposeful human tool reflected the emergence of the new field of functional psychology, a unique contribution of the Chicago School of research in the early 1900s, and, more important, it revealed the growing influence of John Dewey, who was to take over from G. Stanley Hall as the leader of the educational reform movement in the United States.

The shift from Hall to Dewey happened in the first decade of the new century. As influential as Hall's ideas were in the 1890s—and influential they were in prompting educators to consider the learning process from the perspective of the pupil rather than that of the disciplinary expert—Hall fell out of favor both within the field of psychology, which moved to adopt more experimental methods of study, and in education, which looked to scholars whose ideas mapped more productively onto a view of education as a positive force for social change. Hall's romantic notions of childhood and social Darwinist commitments were increasingly poor fits for the role schools were beginning to assume as institutions of individual and social progress, institutions tightly connected to the everyday experiences of people in their communities. Functional psychology was just what educators were looking for to spur on their work.[14]

At the center of functional psychology was a view of human consciousness as a tool or instrument for helping organisms cope with their environment. The emergence of this view of consciousness as an adaptive feature came directly from Darwinian ideas of evolution that were popularized in the late nineteenth century. Rather than seeing the mind as something distinct from the body, it was recognized as a natural extension of the organism. That is, it was understood to be a biological adaptation, just as a bird's wing or a giraffe's neck is. Adherents of this view clustered at the University of Chicago and included faculty such as the social psychologist George Herbert Mead, psychologist James Rowland Angell, philosopher James H. Tufts, and John Dewey. This view of consciousness, or thought, which Dewey termed "instrumentalism," was extended and applied to a wide variety of settings in a burst of intellectual activity from the mid-1890s through the first decade of the twentieth century.[15]

The Chicago School of thought in psychology emerged from a more extensive intellectual commitment to evolutionary ideas held by university faculty in this period. Many believed that the primary characteristic of all life was its tendency to continually change in ways that increased its fit with its surroundings. This idea of "progressive evolutionism" infused the work of researchers such as Jacques Loeb and Charles Manning Child at Chicago, who studied interactions between organisms and their surroundings. Living things, they were convinced, did not passively receive stimuli from the environment. Rather, each and every organism throughout its life worked actively to alter its habitat to meet its particular needs, and the process of doing so directed its development along a natural positive trajectory.

Such progressive ideas colored much of the scholarly work in the founding years at Chicago, and they moved easily from one disciplinary field to the next. One could find researchers in fields as disparate as psychology, biology, physics, education, and philosophy drawing from this common pool of progressive thought as they went about their work. The extension of these ideas beyond the strictly biological to social and cultural phenomena gave rise to functional psychology, and the implications, put simply, were that intelligence could be developed and had the power to improve the human condition. Some advocates went so far as to argue that the "scientific investigator" and science as a form of "intellectual activity" were themselves both products of evolutionary development as well as the "primary agent" of that development. A natural consequence of this commitment was for scholars to become actively engaged in the pressing affairs of their time. Dewey led by example.[16]

Dewey earned his PhD in philosophy studying with George Sylvester Morris at Johns Hopkins in 1884, though his interests included education and psychology as well. Before his graduate work with Morris, he spent three

Figure 8. John Dewey, ca. 1925 (from a photograph by Keystone View Co., New York). The New York Public Library / Art Resource, NY / ART194616.

years teaching high school, two in Pennsylvania followed by a year in Vermont. During his time at Hopkins he studied physiological psychology under G. Stanley Hall, who, along with William James, was one of the key figures of the new psychology in America. Dewey's first faculty position was as an assistant professor in the philosophy department at the University of Michigan, a department he later took over as chair after a brief stint on the faculty at the University of Minnesota. It wasn't until he was recruited as professor of philosophy and pedagogy and director of the School of Education at Chicago in 1894, however, that his work turned strongly toward social reform.[17]

Schools provided natural sites where progressive ideas of social reconstruction might be tested, particularly in the growing urban metropolis of Chicago with its soaring enrollments and large population of immigrant children. In 1896 Dewey (caught up in the laboratory enthusiasm of the time) established his renowned laboratory school for this very purpose; in his proposal for the new school, he even suggested that the field of pedagogy could be thought of as the laboratory where the theories of psychology and ethics might best be studied. Students in this experimental elementary school were taught to see how knowledge could be used as a tool to meet societal needs. Dewey's progressive evolutionist outlook was evident in public remarks he made during the first year of the school's existence. "Man's activities," he asserted, "are marked off from those of animals just because he becomes conscious of the power of control on his part, and becomes conscious of the method of subordinating physical forces to himself."[18]

Science was central to the work of the school, not in the sense of a formal subject to be learned but rather in what it provided as a model of thought that allowed one's interactions with the environment to be moved, as he explained, from mere "accident to that of intelligent control." "Hypotheses in control of action," Dewey's shorthand for scientific method, provided the template for purposeful thinking he sought to convey to students. This method had demonstrated its value in altering the material environment to suit human needs. Directing it to the progressive improvement of all facets of the human condition, he argued, was long overdue.[19]

Dewey's innovative pedagogical work, and particularly his views on the progressive power of science, attracted notice among the scientists and science educators in and around Chicago, Mann in particular. Dewey regularly invited eminent scientists on campus to present their work to students at the lab school and even collaborated with them to develop classroom materials for students to use. Among the visitors were geologist Thomas C. Chamberlain, botanist John M. Coulter, physicist Albert Michelson, and physiologist Jacques Loeb. Dewey enjoyed a particularly close relationship with Loeb; their wives shared similar interests, and their children played together frequently during these years in Chicago. These close connections, both personal and professional, not only facilitated the dissemination of Dewey's pedagogical ideas among the scientific community but also influenced his thinking at a crucial point in his own intellectual development.[20]

Dewey had regular interactions with the science education community as well. He participated in meetings of the newly organized Central Association of Science and Mathematics Teachers, which often met on campus or at local Chicago high schools. In 1903 he delivered a keynote address to the association in which he emphasized the importance of understanding the value science possessed as a means of problem solving. Two years later, Mann advanced a similar vision, no doubt influenced by Dewey's thinking, in a paper on the topic of science education reform. "It is gradually becoming clear," he wrote, that "science must be treated as a part of human experience. It must be so closely linked with the interests and problems of the daily life as to become part of it; it must be shown to have arisen for the purpose of meeting human needs."[21] The functionalist perspective shone through clearly.

In 1904 Dewey left for New York's Columbia University but remained active in science education reform. Mann by then knew enough about Dewey's work to recruit him to participate in the New Movement in Physics Teaching in 1906 and, subsequently, to contribute to the symposium on the "purpose and organization of physics teaching in secondary schools" that was published in *School Science and Mathematics.* When Dewey took the stage at the Boston meeting of the AAAS in 1909 as vice president of the new education section, it seemed natural that Mann was there by his side as secretary.[22]

· · ·

Dewey's ideas about the importance of the scientific method were soon to extend well beyond the local confines of Chicago, New York, or even Boston, where the AAAS gathered to hear his address. They were delivered in the pages of *How We Think.* Dewey wrote the book for teachers based on his seven years of work at the laboratory school in Chicago, and in it he provided a remarkably accessible description of the process of science that enabled its easy application to problems of daily life.

All thinking, he explained, was essentially a problem-solving activity. But rather than portray thinking as some combination of observation, hypothesis formation, and varying degrees of induction and deduction, as logicians and scientific methodologists had done for years, Dewey adopted a psychological perspective. Effective thinking, he argued, began with a problem that was resolved through appropriate reflection and consideration of the consequences of the action taken for its resolution. For educators, this problem-solving gloss on scientific method provided a welcome avenue for connecting science with things that excited student interest.

In *How We Think* Dewey sought to provide a solution to the specialization about which so many scientists and educators had grown concerned. Rather than continue to bombard students with ever more facts, which was often done "in the attempt to make the work interesting and practical" but ended up instead overwhelming students and "are soon forgotten," as one educator observed, Dewey wished to focus on teaching a method of thinking. "This book represents," he wrote in the preface, "the conviction that the needed steadying and centralizing factor is found in adopting as the end of endeavor that attitude of mind, that habit of thought, which we call scientific."[23] This would be the means, he believed, to unify and focus teaching in a way that would meet the needs of the wide range of students then in the schools.

The most influential part of the book, at least as it related to science education, was not about science education at all, but rather where Dewey laid out what he called a "complete act of thought." Any such act of thought, he explained, consisted of five "logically distinct" steps: "(i) a felt difficulty; (ii) its location and definition; (iii) suggestion of possible solution; (iv) development by reasoning of the bearing of the suggestion; [and] (v) further observation and experiment leading to its acceptance or rejection."[24] He used three everyday examples to illustrate how the thought process progressed through these steps. In the first, he tells the story of a man finding himself downtown who suddenly realizes he needs to be uptown in forty minutes for an appointment. The man considers the various transportation options available and concludes that the subway is his best bet. The second example involved puzzling out the function of a steering pole on a ferryboat. The closest thing to science is found in the last example, in which an individual who has washed a glass in warm soapy water wonders why the bubbles that formed on the outside of the rim were pulled inside after it was set mouth down on a plate. For each of these situations, he provided a detailed analysis of the thought process involved, breaking it down into its "elementary constituents" (the five steps listed above) and showing how in the end the puzzling situation was resolved. Although Dewey didn't present them as such, these five steps would soon be held as the essential elements of the scientific method for generations of teachers and students to come.

All of Dewey's scholarly and pedagogical work was closely tied to the epistemology of science. His association with scientists during his time at Chicago and elsewhere left its mark, without question. Writing to a colleague not long after the publication of *How We Think,* he explained that he came at his "'method' from several points of view" but primarily from that of the physical scientists. "I saw they had a method that worked and so studied the logic of the experimental method . . . hypothesis in control of action." His goal in doing this, however, was not to provide a stepwise account of how

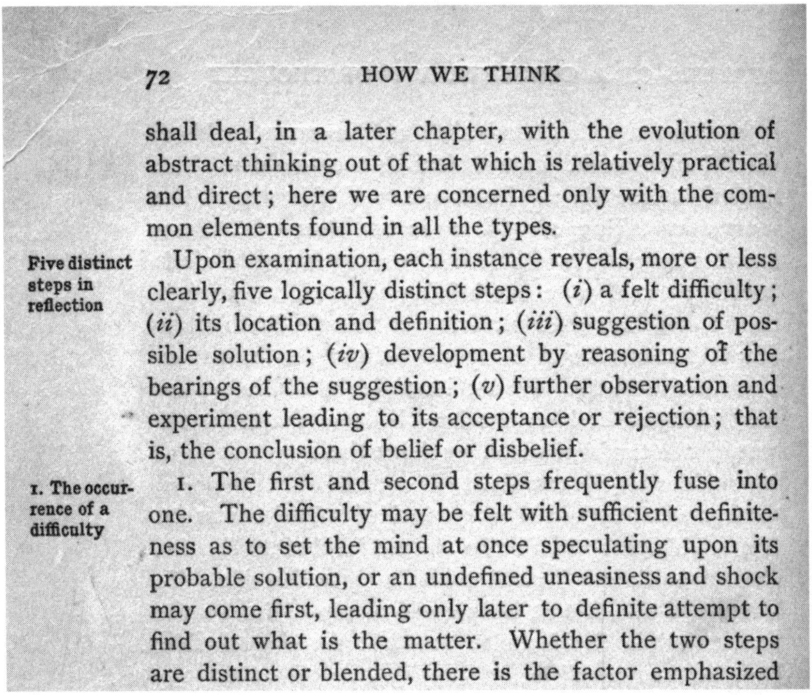

72 HOW WE THINK

shall deal, in a later chapter, with the evolution of abstract thinking out of that which is relatively practical and direct; here we are concerned only with the common elements found in all the types.

Five distinct steps in reflection Upon examination, each instance reveals, more or less clearly, five logically distinct steps: (*i*) a felt difficulty; (*ii*) its location and definition; (*iii*) suggestion of possible solution; (*iv*) development by reasoning of the bearings of the suggestion; (*v*) further observation and experiment leading to its acceptance or rejection; that is, the conclusion of belief or disbelief.

1. The occurrence of a difficulty 1. The first and second steps frequently fuse into one. The difficulty may be felt with sufficient definiteness as to set the mind at once speculating upon its probable solution, or an undefined uneasiness and shock may come first, leading only later to definite attempt to find out what is the matter. Whether the two steps are distinct or blended, there is the factor emphasized

Figure 9. The five distinct steps in reflection from John Dewey, *How We Think* (Boston: D. C. Heath, 1910), 72. Reproduced from a copy in the author's collection.

scientists went about their own work but rather to use the method of science to provide the framework for reflective thought in everyday use—to account for how thinking was most effectively used to guide practical action, as his examples clearly reflect. As Dewey explained in his Boston address, the "scientific method" was not only thinking "for highly specialized ends; it *is* thinking so far as thought had become conscious of its proper ends and of the equipment indispensable for success in their pursuit." The extension of the scientific method of reasoning—in this psychological framing—to the problems of the everyday world was the bold project to which *How We Think* and nearly all of Dewey's other work was directed.[25]

D. C. Heath and Company published *How We Think* in March 1910. However, Dewey's editor, the former Chicago schools superintendent Edwin Cooley, unexpectedly left his position at Heath right about that same time, casting a pall of uncertainty over the book's fate. Dewey worried the work might get lost in the editorial transition. He complained to Frank Manny, an old friend at the University of Michigan, that the staff at Heath weren't "doing anything especial to push the book" and added that "Mr. Cooley's going out just as he did won't help the book any I suppose." The favorable reception the book eventually received must have been a pleasant surprise under the circumstances. Reviews were positive on balance, and the book became something of a sensation. It went through more than twenty printings before it was revised and republished in 1933. Nearly two decades after its original publication, no work on the Heath list was quoted more frequently.[26] Much of that success was due to Dewey's simplified definition of the "act of thought," which many understood to represent the essence of the scientific method.

Given the close association between the method of science and reflective thought that Dewey held, it is not surprising that readers would inadvertently view his five steps—what in reality was a greatly generalized abstraction of the scientific process—as the scientific method itself. One reviewer insisted that the title of *How We Think* was a "misnomer." Rather than an account of the process of thought, he wrote, the book was "in reality a plea for scientific method in . . . school," and in many ways this was true.[27] What is interesting, though, is not how Dewey used science but rather how educators used Dewey to reconstruct the scientific process for high school students. The willingness of readers, particularly educators, to anoint these steps as *the* scientific method can be attributed to the fact that this personal, cognitive version of method meshed well with the prevailing trend in education toward the psychology of student interest and real-world problems initiated by G. Stanley Hall a decade or so earlier. Dewey's method presented a universal means of approaching any situation from a scientific point of view without having to bother with formal rules of logic. It allowed educators to embrace the rhetoric of science and thus to take advantage of the cultural authority science possessed following the wave of late nineteenth-century popularization, and at the same time it provided a legitimate avenue for bringing real-world problems into the science classroom.

•   •   •

Equating reflective thought, or generalized problem solving, with the scientific method almost immediately resolved the tension classroom teachers were experiencing between teaching about the utility of science, on the one hand, and the broader push to emphasize scientific process, on the other. Leading science educators readily saw the value in the new approach and repeatedly used it to pillory the rigid laboratory method as well as the theory of mental discipline on which it was based. "Recent psychological research has struck so many blows at the old doctrine of formal discipline," wrote one Ohio high school teacher, "that we have come to view it in about the same light as the small boy did the vacuum when he said, 'a vacuum is something with nothing in it.'" Transfer of thinking skills from the laboratory to daily life simply didn't happen. "If we expect our pupils to employ the scientific method in the future," this teacher asked, "why in the name of common sense do we not give them a chance to use it in their high school science work?" John G. Coulter of Indiana University (John M. Coulter's son) added his voice to the attack. In a talk at the meeting of the National Education Association, he noted the significance of Dewey's vision and explained that "the psychologists have punctured the complacent doctrine that we can train in the scientific spirit in the artificial realm of the laboratory and then expect this training to be carried over into the practical relations of successful living." "We propose to teach scientific method," he stated, "as an ideal of general application rather than as a specific kind of skill."[28] The ideal was scientific thought. The psychological steps were Dewey's.

It wasn't long before the "problem method" took on a life of its own as the "latest and best in theory and practice." In a piece written for Chicago schoolteachers, a district administrator highlighted the value of the approach and explained—somewhat pedantically—its foundation in psychology and science. "The problem," he wrote, "calls for a new form of action and thinking is the means of establishing this new form of action. Modern function logic and modern biological psychology coincide in their emphasis on the relation of problem, purpose, and thought." Examples from the sciences, he went on, have "contributed to the theory and practice of the problem method." "It might not be too much to say," he added, fully aware of Dewey's writing, "that science has been the chief source of the movement."[29]

Though not commenting specifically on Dewey's use of scientific thinking as a model for everyday problem solving, well-known scientists of the time seemed to be in general accord with the direction of his work and the more popular trends toward greater student engagement. Chicago physicists Michelson and Millikan clearly supported the new direction. In the midst of the debates over physics education sparked by Mann's New Movement in Physics Teaching, Michelson voiced his approval, and in the same spirit

Millikan welcomed the kind of teaching that would exist "for the sake of meeting . . . the present and future needs of the pupil"—"both for the sake of his own happiness and the sake of his usefulness in the community." In a presentation before the AAAS a year after Dewey's address, Harvard anatomist Charles Minot made a notable shift, downplaying the value of logic in understanding science and adopting a functionalist perspective in its place. "It must be doubted very seriously whether the study of logic is really essential," he said. "'The method of science' means more than logic"; rather, the phrase referred to "a complex physiological function, which is brought to efficiency by practice."[30]

Others pushed for the wider application of scientific methods to all aspects of human experience, as Dewey did, and some deplored the fact that "a few scientific men should persist in interpreting scientific method in such a way as to limit its application to purely physical phenomena." Dexter Kimball, a professor of engineering at Cornell, explicitly commented on the changing view of method in the pages of *Science* in 1913: "The term 'scientific method,' has come to mean a somewhat definite way of approaching the solution to all problems as opposed to older so-called empirical methods."[31] This conception of scientific method broadly expanded its scope of application, making it something much more immediately useful to the average person on the street. But a direct consequence of its expanded power in everyday affairs was the marginalization of disciplinary knowledge and scientific expertise. What need did the public have for scientists or scientific judgment if anyone could simply learn to follow the predetermined steps of the scientific method? This, as we will see, would emerge as a serious concern in the decades to come.

The shifting view of method, both in conceptualization and in scope of application, was not the result of Dewey's work alone and certainly was not traceable to his book *How We Think* in an immediate or direct way. These changes were all part of the new pragmatic spirit that infused and shaped the wider cultural configurations of science in the Progressive Era.[32] What Dewey's slim volume did accomplish, however, particularly for educators, was to provide in the five steps of thought a powerfully epitomized statement of scientific process. Its brief and intelligible formulation was the key to its successful dissemination among teachers and other readers.

Reviewers commented repeatedly on the striking accessibility of Dewey's ideas. The Illinois philosopher Boyd Bode wrote that *How We Think* was the "rare kind of book in which simplicity is the outcome of seasoned scholarship." "It is no small thing to have stated the problem clearly enough to be apprehended by the teachers of little culture and narrow horizon," noted another reviewer. Max Eastman, a student of Dewey's at the time, began his review with the five-step complete act of thought, which, he explained, embodied the essence of Dewey's philosophy "in a form and language comprehensible to minds uncorrupted by philosophic scholarship." Another reader declared that the book was "written in Dewey's best style, abounding in picturesque comparisons and concrete examples," and noted that chapter 6 "gives an analysis of a complete act of thought that will become a classic."[33] Indeed it did.

The new view of method was widely disseminated throughout the professional education establishment. Textbooks on how to teach science, relatively rare before 1910, proliferated with the expansion of public schooling and new teacher-education programs. These books often opened with chapters laying out the goals and objectives of science education, which now highlighted student problem solving or its equivalent, the use of the scientific method in everyday life, followed by a description of just what that method was. Textbook authors increasingly invoked Dewey rather than Bacon, Mill, or Pearson on the topic, and where previously they had struggled in their attempts to describe the logical elements of science in a simple, straightforward way, after *How We Think* they could point to the five steps or the central place of the "problem" as the psychological starting point of scientific investigation. Mann, for example, made Dewey's methodological approach the centerpiece of his discussion of scientific method in his 1912 textbook on physics teaching. Another author defined scientific method in this same vein, noting that "the true way to induct beginners into its use" was to present them with problems and then "guide them in using the scientific method in reaching their solutions," which, he explained, consisted of the Deweyan steps of the "complete act of thought."[34]

By 1918, in an interesting turn, some science educators drew on Dewey's account of reflective thinking to illustrate the work of some of the great figures in the history of science, individuals such as Pasteur, Darwin, Galileo, and Newton. Casting the thinking of these scientists in Dewey's terms, in their view, was more powerful pedagogically than adapting the formal elements of method commonly used by academic philosophers. One writer who profiled these giants of science wrote emphatically that "the great evils of science teaching of today are due chiefly to the adherence of science teachers to a false analysis of the method of the scientists. The formal logic steps of Bacon or Mill or some of the other metaphysicians attack the problem from the wrong end, as far as the educator is concerned." "Dewey's analysis of thought which I have attempted to apply to the work of the scientists," he explained, "will solve the problem of science teaching."[35] In this way, the epic stories of scientific progress provided concrete examples of the Deweyan steps of thought, further cementing the association between them in the minds of teachers.

The transformation of the scientific method in schools (or at least in the eyes of science educators) from its logical formulation as the interplay of sterile inductive and deductive intellectual moves applied—often implicitly—to springs, pulleys, chemical reactions, or dissected specimens to a dynamic, personal method of solving everyday problems can be attributed to a number of factors. The growth in enrollment that filled existing classrooms and forced the addition of countless others in the new schools being built across the country was without question an important trigger for change. But there were other factors as well that shaped the nature of this transformation. The growing interest in psychology and its application to the learning process provided a viable alternative framework ready at hand that could replace the outmoded ideas of mental discipline from the previous century, and the growing network of professional educators—and (after 1909) scientists themselves—provided an effective means of disseminating to all corners of the educational system this new view of how science was done.

In the decades after the publication of *How We Think,* science education in the United States built heavily on the twin foundations of scientific method and student interest, both of which derived from the new psychological ideas that emerged around the turn of the century. These ideas, steeped in a brew of progressive evolutionism—an ideology committed to the belief that society had the power to engineer its own improvement through the application of scientific thinking to the environment, served to guide teaching, writing, and research in high school science through midcentury. In classrooms, this would take place in the form of projects, a new pedagogical method born of the application of Dewey's problem approach to the practical tasks of daily life.

5 Problems and Projects

THERE WAS, PERHAPS, no better window into how the scientific method combined with the new commitment to student interest in schools than the Children's Science Fair held at the American Museum of Natural History in the 1930s. The fair, launched by the American Institute of the City of New York in 1928, was a grand effort by leading science educators to capture and display the learning outcomes of school science education for the general public. Taking their cue from the county and state fair competitions of rural America, directors of the fair invited city students to develop and submit exhibits that conveyed important ideas about science in modern life. These exhibits, which packed the museum's Education Hall, included things such as "models and charts illustrating principles in physics, astronomy and geology, examples of science in home and city life, and ideas or adaptations of inventions based on scientific principles in the form of apparatus, models, plans or photographs."[1]

The fair was an unqualified success. Submissions the first few years ran into the hundreds, with students participating from schools across Manhattan and the surrounding boroughs. Early estimates pegged the number of visitors in the tens of thousands over the fair's typical week-long run. In 1933 a *New York Times* reporter commented that the fair's success showed that science apparently "rank[ed] second only to athletics in appeal to youthful city dwellers." Among the notable displays in these early days was "a cross-section of an internal-combustion engine, made of tin cans, that 'works' if a button is pressed," a relief map of the New York City water supply, and a display showing the "human being" in its natural environment (occupants of a tiny model dwelling constructed "to illustrate the evolution of the American home and its modern conveniences"). The fair projects were a tangible display of the problem-based teaching going on in science classrooms of the time, and they received their greatest public exposure in 1939 when they were featured at the New York World's Fair, "Building the World of Tomorrow."[2]

Unlike the science fairs and competitions of the later twentieth century, such as the well-known Westinghouse Science Talent Search, students at the Children's Fairs weren't judged on how well they conducted original scientific investigations or experiments. Showcasing a given student's ability to do science wasn't the primary objective. Rather, within the halls of the museum, these students assumed the role of public educator, using their displays to teach visitors how scientific thinking was improving the world around them—the human habitat. Scientific thinking and student interest were the core premises on which every exhibit was built, and the exhibits were intended to epitomize these twin goals of the progressive science teaching of the time.[3]

The first steps down the path toward these displays showcasing the power of scientific thinking to remake the world began in high schools during the 1910s with the adoption of the scientific method as abstracted from Dewey's psychological work. This went hand in hand with a growing commitment to relevance and practical application in the curriculum. Along the way, school science was itself remade, most notably by the nearly simultaneous introduction of "general science" (an entirely new school subject designed to serve as the vehicle for student problem solving) and the "project method" of instruction. These curricular and pedagogical initiatives offered a new definition of what it meant to learn the process of science. Through the lens of "general" science, science operated as a way of thinking—of solving problems—apart from the formal disciplinary content of the specialized sciences. This opened up the potential for its more wide-ranging application to solving problems in the modern, increasingly urban society of the first half of the twentieth century.

●　●　●

The turn toward student interest and Dewey's problem-solving approach resulted in a profound reconfiguration of the science curriculum that was to

Figure 10. Two views of science fairs held in New York City in the 1930s: The Children's Science Fair held in New York City, May 1938, and a close-up of boys viewing the Radio and Airplane Exhibit at the Children's Fair, December 1930. Images # 289546 and 313559, American Museum of Natural History Library.

last through World War II. The immediate effect was to move from the discipline-focused laboratory method to a problem-based engagement with various parts of the student's home, community, and industrial environment. For teachers, this meant that rather than having to choose between teaching the "scientific method" (which before Dewey meant sterile laboratory work) and teaching the practical applications of a given science (which teachers believed were much more likely to excite student interest), they could now do both. These were heady times for high school leaders, who were taking control of the form and function of schooling for the new population of students arriving at their doors. "We are come to the Rubicon," exclaimed an Illinois principal. "We have by rather slow processes of educational evolution reached a point where we, as representatives of the secondary schools of America, must proclaim *autonomy* for the public high school"—autonomy from the overspecialized laboratory approach of the colleges.[4]

In the physical sciences, changes were first ushered in with the New Movement in Physics Teaching. Charles Mann, the force behind the movement, published a new textbook in 1905 along with George Ransom Twiss (a collaborator from Ohio) that sought to provide the spark of learning that was missing from the dry experience of the Harvard course. Mann and Twiss wanted to vitalize physics by demonstrating its connection to the everyday phenomena of life. "The aim has been to show the student," they wrote in the preface, "that knowledge of physics enables him to answer many of the questions over which he has puzzled long in vain." Highlighting the early popular approaches of the British scientists Michael Faraday and John Tyndall, Mann and Twiss filled the book with locomotives, flywheels, steam turbines, roller-coasters, and other devices of the era in an effort to excite students with "the go of things," as Hall put it so many times. Anticipating Dewey's problem-based formulation, they sought also to demonstrate the scientific method by leading with inventions or general phenomena, which "may not be the logical order, but . . . is more nearly the order in which Nature herself teaches."[5]

As the new approach gained traction, the teaching of physics and chemistry in high schools followed suit, adopting a strong practical cast, at least as reflected in textbooks and professional journals of the time. "Physics does not exist in the schools for the purpose of familiarizing young people with either the laws or theories of physics," proclaimed Mann in 1913, "but rather for the sake of helping the pupils to increase their powers of controlling their physical environment intelligently and solving their life's problems rationally."[6]

True to the "controlling the physical environment" part of this vision, the new textbooks of the era overflowed with applied examples and pictures showing that key scientific principles were at work wherever the student looked. Charles Dull leavened the pages of his inaugural 1922 edition of *Principles of Modern Physics,* a textbook that came to dominate the high school market through the middle of the century, with vacuum cleaners, hydroplanes, phonographs, and battleships (this last example being a nod to the recently concluded Great War, which other authors jumped to include as well—no doubt for the excitement it would stimulate in students). Carhart and Chute's 1927 *Practical Physics* similarly brimmed with automobile chassis, steam hammers, and submarines. Chemistry textbooks paralleled their physics counterparts, detailing the chemical processes of steel production, glassmaking, and bleach preparation. They examined the operation of acetylene torches, fire extinguishers, and gas masks used during the war, among numerous other chemically related appliances and inventions. The message they sought to convey was that life was easier and America could do more because of science.[7]

Though many of the authors highlighted the importance of student problem solving in their prefatory comments, as Mann did in his 1905 textbook, explicit treatment of how science worked as a problem-solving endeavor was largely missing from these upper-level textbooks (physics and chemistry were typically taken in the later years of high school). The extent to which students actually encountered problems to solve was limited to the questions listed at the end of each chapter. There the authors would ask students to consider things such as "Why do human voices differ when singing notes of the same pitch?" following a unit on sound or whether "it is economical to have steam radiators highly polished" after learning about heat. One 1932 physics textbook offered a section of "For Investigation" suggestions after each chapter that had students find out, for example, how "pneumatic-tube systems in department stores work" or asked them to "set up a simple telegraph circuit and show how it works." The problems were more "finding out" about the science behind some everyday technology than they were about having students grapple with some genuinely puzzling natural phenomenon.[8]

More significant changes came in the life sciences as the subject of
biology emerged around 1910 from the fusion of the individual subjects
of botany, zoology, and physiology (the last of which was introduced in response to state temperance laws that required education about the evils of
alcohol). The transformation was dramatic—in terms of both content and
method—and was well captured in a talk by the New York biology teacher
and textbook author Henry Linville at the same Boston AAAS meeting in
1909 where Dewey gave his landmark address. In his talk, Linville compared the old and new approaches to biology education. He explained how,
until then, zoology and botany had been limited to the morphological
study of type organisms with the goal of demonstrating the "order of evolution in organic development," and that this study was primarily pursued
through the laboratory method popular at the time. The result was that
"thousands of young, untrained pupils were required to cut, section, examine, and draw the parts of the dead bodies of unknown and unheard-of
animals and plants, and later to reproduce in examinations what they remembered of the facts they had seen." Never, he said, in doing all this did
we give "any thought to human interests in connection with our work."[9]

There was the belief that somehow such study would be useful to the student outside of the classroom; no doubt the laboratory work would cultivate
the student's intellectual "powers," as so many had claimed over the years.
Linville told the story of a morphologist who had insisted to him that training
a boy "to solve the problems of the structure of the crayfish" would enable
him "to help in the solution of important industrial and financial problems
in life." This was a stretch, as Linville and those listening well knew. Why
not take a more direct approach? he argued. The "crying need of the times
is for men and women who think clearly. The great and splendid opportunity of all the sciences in the schools," he exclaimed, "appears to me to be
this: They may use the method of the experiment in school on problems that
come up in life itself."[10]

A biology course organized around the problems of life looked radically
different from one organized around assorted type specimens and their evolutionary relationships. It would include, according to Linville, topics such
as the biological importance of food, its preparation, and its digestion; sani-
tation; sex and hygiene; the organic causes of disease; and the conservation
of natural resources. "Most biologists will say that this is not biology," he

noted, and he agreed: "It is *not* the subject matter of conventional biology, but it is the subject-matter of life."[11]

One of the first and most popular general biology courses was that offered by George Hunter in his 1914 textbook *A Civic Biology, Presented in Problems*. The problem approach, evident from the title, was utilized throughout, and the main goal was to promote in students the capacity to think and solve problems in their immediate environment. "The boy or girl of average ability upon admission to secondary school is not a thinking individual," Hunter explained in the foreword. "Therefore, the first science of secondary school, elementary biology, should be primarily the vehicle by which the child is taught to solve problems and to think straight in doing so." The book essentially followed the outline offered by Linville in his Boston address. It covered the relationship of humans to their environment, the role of nutrition in plants and animals, the economic importance of animals, the "human machine" and its needs, and so on. Each chapter began first with the problem to be addressed, such as in chapter 2, on the environment, where the problem was "to discover some of the factors of the environment of plants and animals." This was followed by suggested laboratory activities or demonstrations (in this case for students to view animals and plants in a vivarium) as well as a home exercise (students were to "study the factors making up my own environment and how I can aid in their control"). Other textbooks of the period followed this general model, all placing a heavy emphasis on connections to humans and their everyday needs, but they varied in how they framed tasks as "problems" for students to engage, at least early on.[12]

The problem focus came to the fore more strongly in the 1920s and 1930s. Textbook authors, moreover, began to discuss explicitly the process of science as a problem-solving endeavor. Fitzpatrick and Horton's 1937 *Biology* led off the end-of-chapter activity list with some aspect of the scientific method, such as "a scientist analyzes problems and objects into parts," that students would then have the opportunity to practice. The late 1930s version of Hunter's textbook was renamed *Problems in Biology* and included, in addition to lists of "practical problems" at the end of each chapter, a long introductory discussion on the nature of scientific work in biology. On the value of experiment, Hunter framed it as the psychological method of that "great thinker and writer, John Dewey," explaining that Dewey "has likened this method of problem solving to the steps in the act of thought." He then proceeded to list the steps, beginning with the problematic situation and ending with the "answer to our problem."[13]

The transformation of physics and chemistry and the emergence of biology during these years was enabled without question by the work of Dewey and psychologists such as G. Stanley Hall. Both came to their new ideas about the form and function of education as a result of what they saw as declining public interest in science. The concern Dewey voiced at the Boston meeting about students not flocking to the sciences in the numbers predicted was really just the tip of the iceberg. Indeed, the U.S. commissioner of education, while noting in 1912 that the rise in high school enrollments "constitutes one of the remarkable features of educational progress in the first years of the new century," noted as well shrinking enrollments in the sciences. "One can see that Latin is holding its ground; Greek is disappearing; French and German are gaining," but "all the older sciences, rather strangely, are relatively falling off."[14]

These downward trends had been evident for some time, in fact, and in a series of articles and addresses science educators debated the reasons for the apparent lack of student interest. They expressed their frustration that more students weren't drawn to the subjects, particularly in what was widely held to be an age of science and technology. One concerned teacher extrapolated with alarm that at the "existing rate of decline physiology will cease to be studied by 1925, physics by 1935, [and] chemistry by 1945."[15] Physiology as a separate subject did disappear, in fact, with the coming of the new biology. Chemistry and physics, for their part, moved to connect their subject matters with students' real-world interests in an effort to stem their continued slide. The most radical change, however, was the development of an entirely new course, general science, which many hoped would not only draw student interest (and larger enrollments) but also be the vehicle for disseminating the scientific method among the masses of students who now attended high school.

•   •   •

The first general science courses appeared in scattered locations across the East and Midwest around 1906. These adopted a variety of organizational formats, including everything from a course made up of equal parts elementary physics, chemistry, and biology to courses based primarily on a modified form of physiography (physical geography) or introductory physics. More thoroughly integrated courses began to appear by the end of the decade. All were designed from the outset, in the words of an Ohio teacher, "to give new treatment to the old subjects and to enrich and enliven them by emphasizing the attractive or even spectacular features" of science. In considering the beginning high school student, another advocate of the new course explained, "it is important that his first impressions of this 'wonderland' be favorable, that he be happily introduced into its mysteries and methods, and that the fire of his youthful eagerness be not quenched." "Is this not indeed the time," he asked, "to open wide the door and let the student look in at the many good things that are ahead and to start him happily on the way to acquiring the methods of science?"[16] This was one of the primary objectives of the movement toward general science, and the method underlying it all was Dewey's.

The course that had the greatest impact on setting the pattern for all general science courses to come was developed in 1910 at Chicago's University High School, the secondary counterpart to Dewey's elementary-level laboratory school. Even after Dewey left for Columbia University in 1904, the intellectual atmosphere of the University of Chicago continued to nurture educational innovation, this time giving rise to general science. The architects of that offering were Otis Caldwell and William Eikenberry. Caldwell, who had obtained his PhD in botany at Chicago, took over as head of the natural science department in the School of Education at the university in 1907 and from that position oversaw science instruction at the university high school. Eikenberry was a career science teacher with a zeal for curriculum reform whom Caldwell recruited from St. Louis.[17] Together they represented the new breed of progressive science educators—individuals rigorously trained in the sciences, but committed to a more socially relevant kind of science teaching built upon the new psychological ideas of the time.

Caldwell was concerned by the narrow disciplinary work he was seeing in the science offerings at the school. "Specialists in science," by which he meant the research scientists of the time, had become "so highly specialized that they did not know the fields of other specialists." And he realized that, given their focus on disciplinary advancement, they "would not make much contribution to the problem of science education," which, from his perspective, was a generalist's game through and through. It wasn't that the research scientists at Chicago weren't interested in helping out; Caldwell conceded that he did indeed have "the cooperation of some of the ablest of university scientists" (Robert Millikan participated, for example). But if a new vision of science teaching was to be realized, he thought, it would need to come from educators like him.[18]

Caldwell and Eikenberry developed the new general science course with help from Mary Blount, a science teacher at the school who had recently completed a PhD in zoology. It was, by design, a "synthetic course," one in which the boundaries between the sciences were "as inconspicuous as possible." As Caldwell explained, rather than have some scientific discipline provide the framework for the curriculum, "the topic would be the unit of the course." This was the only plan "that can give us what may really be called a course in general science." In taking up each topic, "it is true," he granted, "that we may be led well out into a tributary science but from it we come back to the main stream and continue along its wider course soon again to be led out along another tributary, again to return." "What is the main course?" he asked. "It is those things in science that present to first-year pupils in a most elementary way appreciable problems dealing with the earth and its processes, plants and animals as they are most closely related to near-at-hand and common experiences of men." Caldwell felt that the University of Chicago research scientists who helped out, some of whom had children in the school, came to see the benefits of the general approach. As he put it, they "were sorry for their specialty but glad for their children."[19]

The defining feature of the course—more so even than its focus on topics of student interest—was the problem method of instruction. Beginning with some familiar natural phenomenon, "the pupil is brought face to face with the problem which in the existing state of his knowledge he cannot solve." The problematic situation thus provided the motivational context for the use of the scientific method. "As soon as the problem is properly developed," Eikenberry explained, "the pupil is guided in making a sort of side excursion in quest of the necessary information, later to return to the solution of the original problem." For Caldwell and Eikenberry, this was the primary goal. Responding to questions about whether the course had any scientific value, Caldwell replied, "It would seem that the scientific value should be enhanced by the fact that thinking has been done about things of common life, things about which the pupils must continue to think about when out of school." This was the whole point, he explained: "The scientific method can never be of largest good until people get to using it in daily life."[20]

After a few years, Caldwell and Eikenberry gathered their materials and published them in 1914 as a textbook, *Elements of General Science,* with the Boston house Ginn and Company. They organized the content of the textbook around five primary natural phenomena—the air, water and its uses, work and energy, the earth's crust, and life on earth. Each chapter covered the various scientific understandings of the topic in question and often described experiments that could be done to demonstrate the underlying principle involved. Rather than listing problems and questions at the end of each chapter, Caldwell and Eikenberry had them printed in a separate laboratory manual. These asked students, for example, "What is the relation between water supply and disease?"—a question that could be answered by referring to the data presented in the textbook on mortality rates from typhoid fever in different cities around the world and comparing those to the manner in which those cities obtained their drinking water. All the laboratory exercises shared this same format: students were provided a problem and then given directions about how to either find data in the text or perform some hands-on activity to generate their own data in order to solve it.[21]

Elements of General Science was one of the first in a long line of textbooks aimed at the first-year general science market. The problem-solving framework of these books enabled educators to dispense with discipline-based science content altogether. After all, the specialized knowledge of any one science wasn't what reformers saw as valuable to high school students. Since the 1890s it had been the scientific method (then the laboratory method) that provided the strongest justification for science education. The new subject of general science promised method (via problem solving) first and foremost, and applied it to topics and problems that were actually relevant to students' lives.[22]

The editors at Ginn and Company wasted no time promoting the benefits of the book's approach and, not incidentally, the connection to Dewey's recently published bestseller *How We Think.* A full-page advertisement for Caldwell and Eikenberry's book led off with "To Think and to Know How to Think," and below that it said, "Teach your students *how* to think consistently, clearly, logically." The ad went on, "This is one prime reason why the general science course . . . has met with nation-wide approval. Just as it fills a definite need in the school curriculum, so does—Caldwell and Eikenberry's *Elements of General Science.*"[23]

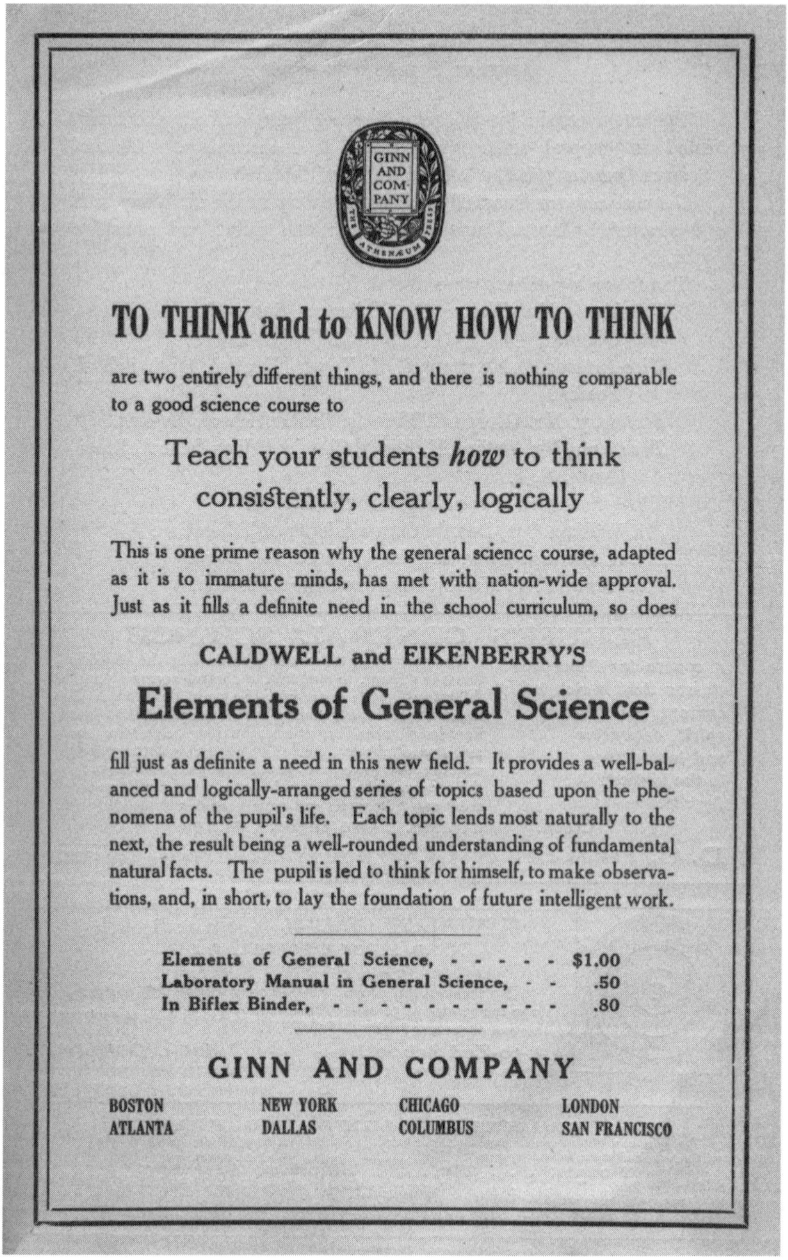

Figure 11. "To Think and Know How to Think," advertisement for Caldwell and
Eikenberry's *Elements of General Science* textbook, from *General Science Quarterly* 1,
no. 1 (November 1916): 63. Reproduced from a copy in the University of Wisconsin
Libraries.

The connection to Dewey's ideas was more than superficial. They provided the foundation for Caldwell and Eikenberry's course as well as for the general science movement as a whole. The two of them, naturally, were well aware of Dewey's work at the laboratory school in Chicago. It was during the peak of the lab school years that they were doing their graduate work at the university, perhaps even lending a hand in science teaching at the school. Dewey, for his part, made his views known to the science education community, giving the keynote address at the third meeting of the Central Association of Science and Mathematics Teachers, held in Chicago in 1903, in which he emphasized the role of meaningful problems in science teaching. And at the 1916 meeting of the National Education Association in New York City, Dewey delivered what one audience member said was "a masterpiece on 'Method in Science Teaching,' in which general science was the main theme."[24]

Some years later, after the general science movement was well under way, Eikenberry took the opportunity to lay out the philosophy behind the course, along with recommendations for how it could best be taught, in *The Teaching of General Science*, a methods textbook for teachers. In the chapter "General Science and Method," Eikenberry opened with an attempt to correct a common misconception that the new course was all about its nonspecialized subject matter—the "general" in general science. That wasn't the novel contribution at all. The course, rather, was all about "a reform in method and point of view." He began: "According to Dewey, there is a fundamental unity of method in all thinking, and scientific thinking differs from empirical thinking principally in the refinement and extension of certain phases of the thinking process." These were the steps listed in Dewey's complete act of thought, which Eikenberry repeated verbatim from *How We Think*. It was this method, he went on, that guided the selection of subject matter—the "topics" Caldwell referred to in his account of the course. Six pages later he took the step so many had already taken, equating the act of thought with the scientific method. "The scientific method," he wrote, "is the same in outline as the general method of thought outlined above," and "general science is to be interpreted as a method of developing the habit of scientific thinking."[25]

The informal nationwide launch of general science as a new school subject came in 1916. The topic generated considerable buzz at the NEA meeting in July. During the last three science sessions (held at Washington Irving High School in Manhattan, just a block off Union Square), an observer reported, "nine-tenths of the discussion provoked was on the subject of general science." It was "the one live issue which interested all teachers alike." The paper Dewey presented topped things off, and the enthusiasm teachers displayed "passed all bounds at this session," as an audience member noted. Later that year came the inaugural issue of *General Science Quarterly,* containing Dewey's address. (Caldwell and Eikenberry were on the advisory board of the new journal.) In that first issue, the editor commented that after scanning the programs of the various science teachers' associations, he was "struck by the fact that unusual attention has been given to general science," more attention, in fact, than to the special sciences for which the various associations were organized. Educators that same year convened two general science conferences in New England, which led to the formation of a regional general science club with thirty-seven charter members. It appeared that high school teachers and administrators were actively seeking what the new course offered.[26]

Despite the warm embrace from teachers, the subject was not without its detractors. There were some—scientists, not surprisingly—who felt general science did a disservice to the field to which they had devoted their lives. Most objected to the lack of disciplinary coherence in the course. John M. Coulter, the Chicago botanist who was well acquainted with the new subject (he had been Caldwell's PhD advisor), couldn't abide the topical nature of the work. "A mosaic made up of fragments of information," he complained, "breaks up all natural connections and forbids the development of those ideas which relate and hold facts." A zoologist from Cincinnati, similarly unimpressed, told a gathering of school principals and teachers that "this seems to be the age of fads." "We have our cubists and futurists in education as well as in art," he said, referring to the modernist art movement of the time. Drawing out the analogy, he went on, "It is well within the limits of possibility to suppose that the kind of art that is capable of producing a picture supposed to be that of a man coming downstairs but which in reality resembles nothing so much as a snap-shot of an explosion in a shingle factory, has much in common with the system of education that produces the 'general science' course, by which I mean a year-course dealing with a kind of mosaic of fragments taken from the fields of chemistry, physics and biology."[27]

Much of this discontent stemmed from the educators' deliberate extraction of the scientific method from its moorings in the conceptual knowledge of the disciplines. In framing the scientific method as problem solving, general science advocates pushed the systematized knowledge of the academic sciences into the background, to be drawn upon only as needed in addressing whatever problem students might happen upon. It was this that critics found so troubling. The critic of cubism from Cincinnati complained that "there is too much loose talk about scientific method" and insisted that "method and material must go hand in hand," making the point about the importance of disciplinary knowledge. Millikan, despite his early contributions to the course, doubted whether general science teachers could teach the scientific method at all. "A man cannot inspire in the high school pupil and pass on to him the spirit and the method of science," he stated, "unless he himself understands that spirit and himself knows how to use it. Where is that understanding to be obtained?" he asked. "Not by talking about scientific method, but by getting in close touch with science itself," which meant studying the specialized disciplines.[28]

Such attacks, however, did little to slow the building wave of reform. Following the publication of Caldwell and Eikenberry's book and the excitement of 1916, others jumped into the mix, putting out their own textbook versions of general science. By 1928 a researcher could write: "Today, the field of General Science textbooks constitutes a land of plenty," observing that in the ten years since 1915, twenty new general science textbooks had been published. Another report found that over the five-year period between 1915 and 1920, enrollments in general science courses in the state of Minnesota increased well over 800 percent. Data published by the U.S. Office of Education covering the 1927–1928 school year showed that every state in the Union as well as Guam, the Philippines, and Puerto Rico had high schools offering the course. More than 500,000 students across the country were taking the subject, the most by far of any science; biology was the next closest, with enrollments of just under 400,000. Noting the excitement that general science generated in students, one California geologist in the Roaring Twenties went so far as to suggest offering the class at the college level. "Were all the freshmen who now dawdle through chemistry, biology or botany" to take such a course, he quipped, even "the 'flappers,' would, I believe, come to see the fascination in learning of hydrogen stars, cathode rays, the great seed-bearing fern-like trees of the Carboniferous, etc."[29]

*　　*　　*

As radical as the development and rise of general science was in the school science curriculum, there was perhaps no more radical instructional innovation in the twentieth century than the project method. It rivaled the laboratory method of the nineteenth century for its ability to capture the excitement of teachers and remake the classroom experience of students in schools. Originally developed in the field of agricultural education, it was brought over by high school science teachers around the same time general science got its start. Indeed, given the focus on problems in the general science course, the project method became a natural extension of the desire to marry everyday activities with student-focused work. "General science is project science" was a common refrain.[30] Eventually (in many ways following the path of the laboratory method during its heyday) the project method soon found advocates outside of science in nearly every subject of the school curriculum. It promised a means to engage students in the world around them in a way that the traditional, textbook-based curriculum of the past was never able to do.

The project method emerged as part of the country life movement in America in the late 1800s. The nation's industrialization and urbanization around the turn of the century had drawn large numbers of people to the cities for jobs and a better quality of life, which raised fears among national leaders about the eroding agrarian foundation of the country. School districts, primarily in the Midwest, were among the first to address the issue with the establishment of co-curricular clubs in the early 1900s for students to learn best agricultural practices, in the hopes of increasing student investment in rural life. Boys joined corn clubs in which they had contests to see who could produce the greatest yield from an acre of land, and there were canning clubs for girls, where they would plant, cultivate, and then can tomatoes. The clubs were popular. Advocates were particularly impressed with the pedagogical approach. "The boys are 'learning by doing.' Instead of studying textbooks on agriculture," stated the author of one report, with a nod to the "new education" of the time. The clubs were effective, as he saw it, because "the reality of the process and result grips the growing boy and girl; . . . the activity is actual, the standards are actual, and the results— economic and moral—genuine."[31]

Right around this time, during the height of the reform movements in physics and general science, educators were looking for new ways to engage students in more meaningful learning. Dewey's problem approach was gaining converts by the day. However, in 1914 Mann, in describing a vision of what he called "industrial science," seized upon the agricultural work as a model for what might be done in science as well. He called for teachers to come out from behind their formal syllabi and convey to students their personal excitement for the spirit of science. "It is vastly more fun for the teacher," he wrote, "if he will just be himself and let his enthusiasm spread throughout the class." The best way to do this, in his view, was to set the students on some local task or project, and the best example of this, he noted, was found in "the work of the corn clubs and canning clubs of the south." That same year New York physics educator John Woodhull made a similar call, insisting that "public needs unmistakably require a new organization of science instruction according to *projects*."[32]

The project method was a natural fusion of Dewey's problem-based approach with the new focus on everyday problems. The emphasis that G. Stanley Hall placed on the various technologies of the day—the dynamos, telegraphs, iron bridges, and locomotives—in his early critiques of science teaching were quickly picked up by science education reformers and made the objects of the problem method of study, as was evident in the new textbooks coming out in biology, chemistry, physics, and general science. The industrialization of the United States, though responsible for widespread social and economic disruptions, had created wondrous objects of fascination. It was but a small step to the project method, where the applied tasks of the agricultural clubs could be translated to the bustling urban environment where most of the new high school students lived. Publishers soon brought out separate manuals of student projects for existing textbooks or published books filled with end-of-chapter projects, which included everything from how to secure pure air in a student's bedroom to ensuring the proper operation of a home furnace.[33]

Advocates of the project method were quick to point out its theoretical justification in Dewey's psychology of learning. "It is difficult for teachers to get started with the project method because it is so different from the one by which we were ourselves taught," wrote Mann in 1916. However, "Professor Dewey has given us a formula which is very valuable if it is used intelligently and not too blindly." "The old system of instruction," he explained, "had for its aim, first the development of technique on the ground that this technique might at some future time be of use to the student. The project method teaches technique only in response to the personal need felt by the student himself." One high school teacher wrote that *How We Think* was, in fact, the perfect manual for such teaching. In that book, "Professor Dewey has given us the key for good teaching, and his outline of the process furnishes the method for handling future projects."[34]

Using student projects as the core of science instruction was well matched to the progressive evolutionist ideas that Chicago academics had cultivated in these years. Science, for these reformers, was the tool to be used to re-shape the environment for the betterment of the human condition, and the concrete tasks undertaken as part of project teaching would give students real-world practice in such work. "Coordination with the environment for self-protection and improvement is the life problem which the child faces," a teacher from Wisconsin wrote, and science "offers him help to see and solve his problem." "This is an age of great industrial progress," he continued, "the age of steel and electricity, of mechanical wonders and scientific miracles. Science and service have become inseparable; it is a big factor in the survival of the fittest." The idea was expressed in more grandiose terms by the St. Louis teacher educator Mendel Branom, who wrote, "It is through the project method that the hopes of the civilized world are realized." It was, as he told it, a unique human adaptation. "Man develops," he explained, "not only through the instinctive method, which is characteristic of the growth of all forms of animal life, but also through the project method, which is particu-larly reserved to man."[35]

The emphasis on using projects in science teaching in the progressive evo-lutionist vein—particularly in biology and general science—led some authors to suggest fusing both courses into a new course in "environmental science" with a focus on "environmental mastery," typically of the urban setting in which students lived. This imagined course, more about engineering the cityscape to meet human needs than anything else, was a far cry from the en-vironmental science courses that would later emerge in the 1960s and 1970s, which emphasized the preservation of pristine lands along with the restora-tion of nature spoiled by human development.[36]

The circularity of attribution among the many pedagogical methods and techniques was in some ways dizzying. Dewey, for his part, modeled his complete act of thought—often referred to as the problem method—on the methods of reasoning he observed in his science colleagues at the Uni-versity of Chicago, which effectively moved the scientific method from the esoteric realm of formal research to the everyday world of the student. Ap-plied to the concrete tasks of managing the urban environment, it became the project method, which was then taken by science educators to represent,

II. WATER IN THE HOME

Voluntary Project 1.—To study the plumbing.—Note to the Teacher: This may be made a class project if the school building is supplied by plumbing with hot and cold water. The school janitor or the school engineer will probably help guide the class through the mazes of the basement. After the pupil has been conducted through the school building, he may make the same study of his own home.

Locate the pipe that delivers water from the city mains to the building. Where is the valve that is provided to shut off the water from the building in case a pipe bursts or repairs are to be made to the plumbing? What other provisions are there for shutting off a portion of the water supply without cutting it off from the entire building? Trace the hot and cold water pipes in the basement.

Review Experiments 44, 45, and 52. Then study Figure 93 and explain how the water enters the water heater, is heated, and is forced through the supply pipes when a faucet is opened.

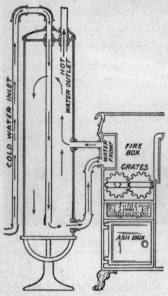

FIGURE 93.—HOT-WATER TANK

Obtain old faucets of different kinds if possible. Take them apart. Bring the parts to class and explain how to insert a new washer or Fuller ball to prevent faucets from leaking.

Figure 12. Typical textbook project from the 1920s. Voluntary Project I in chapter 10, "Water as the Servant of Man," in William H. Snyder, *General Science* (Boston: Allyn and Bacon, 1925), 246. Reproduced from a copy in the author's collection.

in some cases, the essential work in which scientists themselves engaged. Morris Meister, from Columbia University's Teachers College, for instance, wrote that "the scientists have explored unknown regions and have charted the ways. The curriculum is the map which they have left us; but each child must in a measure live through the experiences of the pioneers and reconstruct for himself their map. Thus we have the justification of the project method of study." "The lives of the scientists are lives full of projects," he enthused, and these "projects should now become the projects of the schoolroom."[37]

Through the 1920s and 1930s, student projects became less about modeling what scientists did and more about engaging students in an activity that illustrated the power and benefits of science for humankind. Another Teachers College faculty member, William Heard Kilpatrick, seized upon the project method as the key to the wholesale reconstruction of the school curriculum. In 1918 he published an essay titled simply "The Project Method" that soon became a classic in the field of curriculum and instruction. There he argued that the project, which he defined as a "wholehearted purposeful activity proceeding in a social environment," could serve to unify the educational process for students. Given that "the purposeful act is . . . the typical unit of the worthy life in a democratic society, so also should it be made the typical unit of school procedure" across all subjects. He went on to offer four types of legitimate school projects, only one of which conformed to the Deweyan problem approach to learning, or, as many educators came to see it, the scientific method of problem solving. The rest were little more than things such as "executing a plan" of some sort, such as building a boat, or setting out to attain skill or knowledge in some area, perhaps learning new vocabulary words in a foreign language.[38]

Kilpatrick was a gifted promoter, and his formulation of the project method gained a level of currency among educators that was unmatched in the first half of the twentieth century. At the Children's Science Fair in 1938, it was apparent in the student displays that the idea of a project in science was to "execute a plan" for showing how important some aspect of science was to the improvement of the human condition, more so than to somehow capture and communicate the process of science to the throngs of visitors who came to the museum on Central Park West to learn about science in everyday life.[39]

The difference between the displays at the Children's Science Fair and those at the Columbian Exposition forty-five years earlier is striking. Both presumably sought to showcase the teaching of scientific methods, each in their own way. The Harvard physics apparatus on display in Chicago in 1893, however, reflected a commitment to the scientific discipline, both in its material aspects as well as in its epistemology of precision measurement and inductive reasoning. The displays of 1938, on the other hand, were about science in the abstract, as a tool for human progress. There was little emphasis on disciplinary knowledge; the power of the generalized method to solve problems was everything. Here the methods of science were meant for the individual citizen or for societal use, to be applied outside of the domain of science proper.

Left by the wayside in all this was the school science laboratory. The precision apparatus and laboratory facilities that had been considered essential in the modern high school, and to which boosters had pointed with pride in the late nineteenth century, had by the 1930s and 1940s been deemed superfluous. In his 1934 methods textbook, Hunter asked whether "modern education is justified in making the laboratory the elaborate showroom that it is in some schools." Another science educator was quick to remind his readers that high school laboratories were built "not in response to a need strongly felt" but rather because of "a demand made by the colleges that individual experimentation be done." Educators increasingly saw formal laboratories as working against meaningful science learning. They were overbuilt and overequipped. The *New York Times* reported one science educator saying that "ninety per cent of the laboratory equipment in the elementary and high schools of this country is totally unnecessary." A good teacher, Hunter insisted, needed only common everyday materials and flexible spaces. School districts in the cash-strapped years of the Depression were happy to be off the hook. The laboratory, so highly prized in the late nineteenth century, had become a useless relic in the new age of project teaching.[40]

The radical changes to science teaching and the school science curriculum in this period were part of a broader trend in education toward social utility and the accommodation of student interest that had begun earlier in the century. It was science that led to the new framework for education, and the task of science education was to disseminate this method to the public— through the project method and the new general science course—so that the promise of a new civilization might be realized. Individuals such as Otis Caldwell, C. R. Mann, George W. Hunter, John Woodhull, and George Twiss, among others, led the new progressive wave in science teaching. They worked tirelessly throughout the 1920s and 1930s promoting the cause. The scientific method—the epitome of intelligent thought—was, in their eyes, the key to improving all facets of human existence. "To cause the scientific method to become a part of the bone and sinew of the boys and girls who throng our high schools as to influence their manifold activities throughout their lives," wrote one champion of science teaching, "would be to create within a few generations a new moral and intellectual world."[41] This was the project they hoped to educate the coming generation to undertake.

6 The War on Method

THE BEGINNING OF THE END of the long run enjoyed by the scientific method in schools came in 1945. That year a Harvard committee published the red-bound volume *General Education in a Free Society* (known as the Redbook), which laid out a new vision for education as World War II came to a close. In the portion of the highly publicized report devoted to high school science teaching, the committee, made up of faculty members brought together by Harvard president James Bryant Conant, struggled to reconcile specialized knowledge with the intellectual needs of the nonspecialist (the question that had challenged educators since the turn of the century). Like others before them, they elected to focus on "scientific methodology and the scientific point of view" as the best way to reach the masses. But in their view, teaching "the scientific method" as commonly portrayed in schools was surely not the answer. The authors insisted, in fact, that "nothing could be more stultifying, and, perhaps more important, nothing is further from the procedure of the scientist than a rigorous tabular progression through the supposed 'steps' of the scientific method, with perhaps the further requirement that the student not only memorize but follow this sequence in his attempt to understand natural phenomena."[1]

Other prominent scientists at the time voiced similar dissatisfaction with this view of method. In a talk at the Westinghouse Centennial Forum on "Science and Life in the World" the following year in Pittsburgh, the director of the wartime Office of Scientific Research and Development, Vannevar Bush, described the success of applying scientific techniques to various industrial and logistical problems during the war and called for more scientists to bring these talents to bear on important national problems. "Do not mistake me, however," he added, "as blandly joining the chorus which is bewitched by the magic of the word 'science' and sings an ill-considered and often cloying paean of praise to something summarily referred to as 'the scientific method.'" He went on, "I am certainly not one of those who speak of the scientific method as a firm and clearly defined concept and who regard it as a mystical panacea immediately applicable to any trouble and immediately productive of a complete cure."[2]

The origin of these critiques lay in the dramatic alterations of the social and political climate brought about by the staggering impact of World War II. Science played a central role in the outcome of that conflict, and its place in society was transformed as a result. The use of mathematical models in military operations and the development of new technologies such as radar, the proximity fuse, solid-fuel rockets, and, most awe-inspiring of all, the atomic bomb were key to the Allied victory. Government officials now embraced science as an essential part of the U.S. national security apparatus, and the flow of federal funds, which had begun during the war, continued in order to secure the peace, creating a golden age for researchers. With sure federal backing, scientists were given a free hand to follow their interests, guided only by the fundamental questions of their discipline.[3]

But with the luxuries afforded by government support came new demands and constraints. Key figures within the scientific elite began to recognize that they were not completely free but rather beholden in some form to their federal patrons—and ultimately to the public at large. In the American democratic system, public opinion and public perceptions controlled federal action and ultimately the allocation of resources. More than during any previous time in history, public views of science would be key to the nature of public support and the conditions under which scientists worked.[4]

It was in this context that "the scientific method" came in for attack. Many scientists thought that widespread belief in some singular method of research, as put forward by the science educators during the three decades leading up to World War II, not only was misguided in its failure to capture the true essence of scientific work but also threatened to undermine crucial public support for that work. The scientific method that had been held up as the epitome of rational thinking, as a tool for solving problems in science as well as in everyday life, was now viewed as a clear danger to the scientists'

ATOMIC BOMB CONTINUED

FOR CENTURIES SCIENTISTS HAVE WORKED

SIR ISAAC NEWTON (1642–1727), English, defined fundamental laws of motion ruling atoms as well as larger bodies, revived "atomic theory" that everything is made of tiny particles.

JOHN DALTON (1776–1844), English, converted the vague atomic theory into a scientific asset by atomic weights concept —giving each element a weight, starting with hydrogen's 1.

DMITRI MENDELYEEV (1834–1907), Russian, made the Periodic Table in which the 92 known elements, arranged according to their atomic weights, show a periodic change of properties.

ANTOINE HENRI BECQUEREL (1852–1908), French, while studying X-rays, found by accident that uranium emits invisible radiation and thus, in 1896, discovered radioactivity's existence.

SIR JOSEPH JOHN THOMSON (1856–1940), English, discovered that atoms, which have a neutral charge, contain negatively charged "corpuscles." Later these came to be called "electrons."

LORD ERNEST RUTHERFORD (1871–1937), English, discovered nucleus of atom and first changed one atom into another. However, his transmutations could not be used to produce power.

PIERRE AND MARIE CURIE (1859–1906), (1867–1934), isolated radium and in their studies ascertained that, like uranium, other elements decompose when radioactive rays are emitted.

MAX PLANCK (b. 1858), German, originated the "quantum theory": that energy of radiation is not continuous but exists in small, exact units measured in terms called "quanta."

ALBERT EINSTEIN (b. 1879), German, evolved in a paper on relativity that mass can be converted into energy so completely that no trace is left. Atomic bomb proves truth of theory.

Figure 13. The lionization of scientists in *Life* magazine after World War II, August 20, 1945, 92–93. Reproduced from a copy in the author's collection.

postwar way of life. Its simplicity and potential to be used by anyone on any problem appeared to marginalize scientific expertise itself. As a result, a small group of top scientists had it clearly in their sights. A new battle had begun.

The pointed comments about the scientific method made by Bush and by the Redbook committee were indicative of two primary concerns. For Bush, it was the need to roll back the tendency for the wide-open application of scientific thinking to any and all problems one might encounter outside of the laboratory. It was a call to restrict scientific methods to the phenomena of the natural sciences.[5] The remarks of the Harvard group, speaking in accord with Conant's views on the matter, highlighted the fact that the fight was to be had in the schools and that the target was the oversimplified, five-step method that had been appropriated from John Dewey's *How We Think* decades earlier. Both realized that it clearly mattered how the public believed science operated, and that the task of the scientists in Cambridge was to advance a new, more accurate view of the scientific process that wouldn't threaten the social and political space science now occupied in the postwar world.

•   •   •

The denunciations by Bush and the Harvard committee marked a sharp change in the nature of public proclamations about the value of the scientific method in American society. The progressive science educators going back as far as Charles Riborg Mann and the new movement in physics teaching, of course, had repeatedly made claims for the broad application of scientific thinking to non-science problems. Most students at the high school level, after all, certainly weren't going to end up using science to solve scientific problems, and the transfer studies conducted by the educational psychologists demonstrated that developing such skills in the laboratory had little payoff for students out in the world. Others in the prewar period had advocated for greater utilization of the scientific method in everyday affairs too. Those calls, however, particularly those made by eminent research scientists, largely disappeared after the war as the necessary business of managing the relationship between science and the public began in earnest.

Appeals for the wider application of the scientific method went back as far as the 1870s. There were, as we saw, the pleas made by the naval astronomer and AAAS president Simon Newcomb for liberal training in scientific method to fit nineteenth-century college graduates for the "complex duties" of life. More prominent was the argument for the expansion of method to the social and political sphere made by Karl Pearson in his 1892 bestseller *The Grammar of Science,* which became a touchstone for advocates of the scientific method throughout the first half of the twentieth century. There he insisted that public training in the scientific method was essential to human progress. "The importance of a just appreciation of scientific method is so great," he wrote, "that I think the state may be reasonably called upon to place instruction in pure science within the reach of all its citizens." Such instruction, according to Pearson, would lead to "more efficient citizenship and so to increased social stability." With the continued growth of science and the ongoing work of other advocates of scientific thinking—such as Dewey, who famously insisted that the spread of scientific thought was the "supreme intellectual obligation" of educators—comments like these became increasingly commonplace.[6]

During the 1920s and 1930s, talk of scientific method ramped up even more. Researchers in the social sciences worked to make their fields more "scientific" by consciously using some form of the scientific method in their studies. The Minnesota sociologist F. Stuart Chapin explicitly sought to reconcile Pearson's account of the scientific method from *The Grammar of Science* with research methods in the social sciences in 1924, and around the same time Wisconsin political scientist Arnold Hall was promoting the use of the scientific method in governmental affairs. In that domain, he wrote, "humanity still approximates the jungle type of civilization. Prejudice, superstition and ignorance still hold sway." Clearly worried about the state of international affairs, Columbia president Nicholas Murray Butler publicly insisted on the eve of World War II that "the scientific method must be applied in studying and in solving the political and economic enigmas ahead of us."[7]

The faith in scientific method during these years seemed to be unbounded; some applied it to seemingly anything and everything in the daily matters of life. In the 1920s the *Los Angeles Times* reported on the development in some schools of a "scientific method of musical education." A retailer in New York insisted that with the proper collection of data, it would be possible to have a "scientific method of determining the direction of fashion." There was even a claim by a Reverend Dr. Fleming reported by the *New York Times* in 1932 that "the fact of Christ's birth and immaculate conception" has been "proved by the scientific method used in philosophical thinking."[8] Whether such claims were serious or merely efforts to exploit the cultural authority of science isn't clear.

Of all the public statements about the efficacy of the scientific method before the war, there were none so prominent, however, as those made by the scientists themselves. From his position as president of the California Institute of Technology, which he had assumed in 1921 following his tenure at the University of Chicago, Robert Millikan made repeated claims about its power in the decades leading up to the war. The Nobel Prize he had won in physics in 1923 gave him the public ear and added even greater legitimacy to his methodological pronouncements. "Millikan Sees World Remade by Use of Scientific Method," read the headline of a *New York Times* story in 1930. The following year the paper reported Millikan's belief that "the spread of the scientific method, which has been so profoundly significant for physics, to the solution of our social problems is almost certain to come." At one point he insisted that the country's elected leaders should be more scientific in making their appointments and decisions, and if they did not, they should face remedial work. "A Governor who fails in this should be removed from office. He should be forced to take an elementary course in physics in order to learn the scientific method before running for office again," he was reported to have said in all sincerity. Joining Millikan was MIT president Karl Compton, who insisted during the Depression that more government planning was needed: "Chaos awaits this world unless it adopts and places in effect intelligent planning based on the 'scientific method.'"[9]

By the late 1940s, federal patronage had arrived in full force, changing the overall climate in which science operated. Before the war, the government provided few resources for scientific research (save in areas of agriculture and various smaller, scattered programs). Scientists typically relied on private foundations, their own universities, or research grants from industrial corporations to support their work. During and after the war, however, public dollars for research were abundant, dwarfing all other forms of support. From 1938 to the height of wartime research expenditures in 1945, the U.S. budget for military research and development went from $23 million to over $1.6 billion, seventy times prewar levels. By the late 1940s, funding for military research and development by agencies such as the Defense Department and the Atomic Energy Commission settled in at a still-high thirty times 1938 funding levels—levels that would continue more or less up to the present.[10]

This unprecedented financial commitment to scientific research, in the words of MIT physicist and scientific insider Jerrold Zacharias, "changed the nature of what it [meant] to do science and radically altered the relationship between science and government." Because of this, he explained, "World War II was in many ways a watershed for American science and scientists."[11] In the new postwar world, public understanding of science mattered to the scientific community—not for what science might offer the individual citizen on the street, but for what society should reasonably expect from science and whether it would be considered worth the continuing public investment.

It was under these conditions that members of the scientific elite—those such as Bush and Conant who had helped direct the vast wartime research projects and now mingled with generals, cabinet secretaries, and presidents—raised their concerns about the cavalier manner in which ideas about the "scientific method" had been bandied about. Accepting a method that could be applied seamlessly across disciplines and topics, they worried, had the potential to invite greater external control of the research enterprise. It fed the notion that science was simply a tool that government officials could use to solve whatever problems they deemed needed fixing.

They worried as well about public overconfidence in what science could produce and the notion that application of the "scientific method" could reliably generate immediately useful technologies wherever needed. Decades of high school science textbooks showcasing the technical marvels of science in the hopes of generating student interest had naturally conditioned the public to expect such easy payoffs, and the dramatic, nearly magical products from the war such as radar and the atomic bomb provided ample evidence of this as well. "No allowance is made," however, "for the time-consuming, broadly exploratory work that underlay the . . . successes," noted an official from the American Association of Scientific Workers. "Again science is the victim of an oversimplified image of itself in the public mind."[12]

The concerns over federal control played out in the battle for the establishment of the National Science Foundation (NSF), which began in 1945. Recognizing the importance of science to the national interest, President Roosevelt before his death called on Bush to draft a plan for a permanent foundation to fund basic research in the postwar era. This plan was outlined in Bush's report *Science—The Endless Frontier*. The operation of such a foundation, Bush argued (from the perspective of the scientists), should be overseen by an independent, nongovernmental board to insulate it from overt political control. It was essential that any government support "must leave the internal control of policy, personnel, and the method and scope of the research to the institutions themselves." Members of Congress were equally interested in supporting research after the war but were set on using science to address pressing national problems, and thus they proposed a governance structure that would be more responsive to public needs. After a good deal of political wrangling, the more independent vision of the scientists won out, and in 1950 the National Science Foundation was established as an agency run by scientists for the advancement of basic rather than applied science.[13]

The crucial point at issue in this political struggle over the control of the new foundation was largely methodological. Those who insisted on a structure that allowed for government direction of research viewed science primarily as an intellectual technique, one that could be deployed to solve any problem deemed sufficiently important—be it physical, biological, social, or economic. Bush and other members of the scientific elite, however, put forward a model in which the best science could be achieved only if scientists were free to pursue whatever research questions they found intellectually compelling, without external pressure. Practical applications were inevitable, they insisted, but would emerge only after new basic understandings of the natural world were won, and in no way could be anticipated in advance. Any attempt to direct research to practical ends, according to their argument, would only stifle creativity and impede progress. This view prized the internal, disciplinary knowledge of each of the scientific fields. Physicists knew what the essential questions of physics were, and the same held for biologists, chemists, geologists, and so on. There was no "general" scientific method that could simply be aimed at key national problems, however pleasant it might be "for the majority if they could have all the scientists of the country working for them on any problem which might appear important." As Bush pointed out at the Westinghouse Forum, there was no such thing as a scientific method "immediately applicable to any trouble and immediately productive of a complete cure."[14]

The social sciences proved to be a key area of contention in the founding of the NSF, and here perceptions of scientific method also played a role. Social scientists were gaining influence before the war, and many argued that their field should be treated no differently than the natural sciences in the foundation's funding portfolio. Among the arguments for inclusion was the power such research had to deal with the social consequences engendered by advances in the physical sciences (such as new technologies in industry and nuclear weapons in the realm of national security).

The stronger argument for inclusion, though, was made by those who sought to demonstrate that the natural sciences shared a common method with the social sciences. During congressional hearings on the NSF-authorizing legislation, the education researcher Donald Scates, for example, testified that "techniques have been developed and are widely known for attacking human problems with the same degree of scientific rigor that characterizes physics, biology, and chemistry." "I want to emphasize," he added, "that human science is not basically different from physical science; it simply deals with different phenomena." Others made similar claims, which drew from the unified view of method put forward by Karl Pearson at the turn of the century. But, despite the efforts of researchers such as Scates, concerns about the subversive potential of social science research during the budding Red scare of the late 1940s resulted in the social sciences being left out of the funding picture, at least initially.[15]

● ● ●

The victory the scientific elite won in the establishment of the National Science Foundation was but one battle in a much larger campaign to reshape public perceptions of the process of science. The concerted effort by the scientists to tamp down indiscriminate use of the "scientific method" beginning in the 1940s appeared to have the impact they were hoping for. The use of the phrase in general books, news reports, and assorted public addresses declined significantly in the second half of the twentieth century, to the point where it appeared in newspaper articles primarily as a quaint throwback or symbolic marker for some science-related story.[16] But contrary to its demise in broad public discourse, the phrase and idea continued to thrive in the schools and among the science education establishment, even flourishing in the 1950s and early 1960s. It was there that the scientific research community marshaled more deliberate, systematic efforts to root out scientific-method thinking, particularly in its most reductionist, formulaic incarnations.

The problem-solving approach as scientific method that surfaced initially in the early twentieth century had a continuing presence in most high school science textbooks up through World War II and the 1950s. Though treatments of the scientific method were less explicit in upper-level physics and chemistry books (appearing only in end-of-chapter problem sets for the most part), as one moved toward the earlier high school grades, where biology and general science were taught, detailed discussions of the method were readily apparent. Textbook authors told students in introductory chapters that they "can find scientific problems everywhere" and encouraged students to look for them on their way to and from school, "or wherever you happen to be." Authors explained the value of the scientific method in everyday settings: "In the past, few people could understand scientific methods. Today, science is universal and *scientific attitudes and methods* are not limited to the scientist." And invariably the steps of the method were laid out clearly for the student. "A scientist uses five steps in performing an experiment," one author stated, rehearsing the thoroughly ingrained establishment account of the process.[17]

Even more direct prescriptions concerning scientific method could be found in the various teaching-methods textbooks and education policy reports of the time. All emphasized the close association between problem solving, the scientific method, and the complete act of thought. In 1934, nearly a quarter century after Dewey's book, Hunter offered readers what he called a "simplified statement of the scientific method," commenting that "the fundamentals of problem formulation are so well expressed by Dewey that all teachers know the steps." He went on to quote verbatim those from the frequently bookmarked page of *How We Think*. "The method has been studied and analyzed and its steps are well understood," lectured another author. Such books uniformly justified the method's importance as a learning goal.[18]

Not only did these authors promulgate the five-step method appropriated from Dewey, but some went so far as to advocate the application of the scientific method to the direction and control of scientific work itself—exactly what the postwar scientific elite were fighting against. The authors of the National Society for the Study of Education's *Science Education in American Schools*, published in 1947, for example, wrote, "It may surprise some readers to recognize that a considerable group of scientists regard science as existing chiefly for the benefit of people as a whole. As such, they believe that it should be publicly managed and guided in its endeavors." Such control, they asserted unambiguously, "should come primarily from representatives of the people—from the constituted central authorities and not from organized scientists."[19]

Within this professional sphere of influence, science educators were working with an alternative ideological vision of what science was—an instrumental method of problem solving untethered to the conceptual frameworks or theories of any particular discipline. Moreover, their articulation of that method for use in school science classrooms through the various science textbooks and methods manuals for teachers highlighted its clarity of expression and ease of use. Indeed, the fact that the scientific method they advanced consisted of a series of sequential steps and could be taught to students in isolation from science content knowledge made its dissemination in the schools that much easier. This was especially true for teachers with shaky disciplinary knowledge, a species increasingly common in schools, particularly during the era of exploding school enrollments brought on by the postwar baby boom.[20]

It was the very reasons teachers embraced the stepwise method—its simplicity, ease of teaching, and universal application to any problem—that most troubled the scientists who were paying attention. And the more they looked at how it was being disseminated by educators, the less pleased they were. Back in 1916, when Mann extolled the virtues of Dewey's problem approach, he cautioned that it had the potential to be valuable to teachers if "used intelligently and not too blindly." But between then and the middle decades of the twentieth century, the scientific method had increasingly come to be taught in what seemed to be a blindly rigid and formulaic manner. There was a tendency among science educators, for example, to overspecify the steps in the process. "The teacher who endeavors to direct pupils in acquiring skill in scientific thinking," a Chicago education professor wrote, "must have a clear-cut notion of the elements that constitute such thinking" so that he can "detect and correct the errors that pupils are likely to make in the process." One educator claimed, based on research he had done with scientists defining the steps of the method (he identified ten major and seventeen minor elements), that the high level of agreement he found "encourages the belief that the scientific method has been developed beyond the stage of subjective interpretation and is now becoming recognized as a definite, tried, and tested procedure in problem-solving."[21]

All that was left was to identify the "effective methods and means" of its teaching. For many that method was little more than drill. Hunter in his widely used teaching-methods textbook noted matter-of-factly that "the habit of scientific thinking comes through practice and through the use of specific situations in which the steps of scientific thinking are practiced." Another educator put the matter more bluntly: "The teacher who fails to drill pupils in the scientific method is not doing his whole duty as a teacher of science."[22]

Teaching the process of science in this manner left little room for students to experience the nuances and creativity of scientific work. The very idea of some singular method was antithetical to the complexities of science. "I feel very strongly that we are misleading our fellow-men," said a botanist from Illinois, "indulging in some unwise self-deception by insisting that we have something special—called a scientific method." A fellow scientist from Pennsylvania took issue with what he saw as the common belief in a scientific method as well. Drawing on the typical school portrayals of science, he recognized that there seemed to be widespread acceptance among the public that, as he vividly described it, "science is a sort of intellectual machine, which, when one turns a crank called 'the scientific method,' inevitably grinds out ultimate truth in a series of predictably sequential 'steps,' with complete accuracy and certainty." Others similarly objected to the public perception of the scientific method as some sort of "automatic procedure for solving problems"—"like a steam roller, cracking its problems one by one with even and inexorable force."[23]

Routinized portrayals of the scientific process such as these inevitably (and perhaps deliberately, in an effort to spread the method widely) minimized the creativity, expertise, and tacit knowledge of the scientists themselves, vesting the power of science instead in a simple, disembodied procedure. It isn't surprising that they were met with fierce disapproval from scientists. Shortly after winning the Nobel Prize in physics, Harvard's Percy Bridgman added his voice to the assault in the late 1940s. "It seems to me that there is a good deal of ballyhoo about scientific method," he observed. There were no steps, in his mind: "Scientific method is what working scientists do, not what other people or even they themselves say about it." Bridgman argued that for a scientist, "the essence of the situation is that he is not consciously following any prescribed course of action." If there was a scientific method to be had, he explained, "it is nothing more than doing one's damnedest with one's mind, no holds barred," which was about as far from a "method" as one could get.[24]

•   •   •

Of all the attacks on the scientific method, none was as prominent and sustained as that launched by the Harvard president. Drawing from his first-hand knowledge as a professor of chemistry before assuming his position at the head of the university, Conant took up the call sounded by the Committee on General Education in 1945—a call he himself had helped shape. With support from the Carnegie Corporation of New York, he set out to develop a novel introductory course for freshman and sophomore non-science majors at Harvard that would convey what he thought was a more authentic picture of how science worked. From the mid-1940s through the early 1950s, Conant articulated his curricular vision and—through a series of summer-school courses, conferences, and books aimed at high school teachers, college science instructors, and the general public—sought to correct the rampant public misunderstanding of science. He targeted in particular the idea that the natural sciences advanced by means of a singular, step-by-step method.[25]

Conant was a central player among the midcentury scientific leadership. During World War II he served as chairman of the National Defense Research Committee, an agency that oversaw a variety of research projects that were crucial to the war effort, including among many others the Manhattan Project and one dedicated to the creation of synthetic rubber. His experience working with Washington politicians and military officials in this and other roles led him to realize the crucial importance of public understanding of science. Looking back, he recounted that he and other leading scientists of that time "witnessed the kind of political witches' brew that resulted in those days when technical decisions appeared to be the consequence of the interaction of pressure groups." "Examples of the misunderstanding of science," he declared, "were daily before our eyes."[26]

Public encounters such as these prompted him to create the new introductory course, the pedagogical theory for which he outlined in his 1947 book *On Understanding Science,* which was based on the Terry Lectures he gave at Yale. What was needed, in his view, was for the layperson to develop a true feel for the nature of scientific research rather than learn any particular science content. "The course might be designated a course in the scientific method as illustrated by examples," he explained in an early prospectus. But he balked at any further use of the phrase, believing that "that unnecessary controversies are raised (in the minds of professors if not students) by overemphasis on the words 'science' and 'scientific' method."[27]

Conant's concerns about the scientific method centered on both the pedagogical value of the notion and its fidelity to the actual practice of scientists. He was well aware that the concept needed to be addressed head on in any introductory science course, given the depth to which it had permeated the discourse and culture of the time. And he recognized as well that the strength with which the public clung to the idea was attributable mostly to the success progressive science educators had had infusing it into the school science curriculum. "Indeed, in the last twenty-five years," he explained, "indoctrination in the scientific method has been put forward with more and more insistence as one of the primary aims of modern education." But the "slogan in question," as he put it, troubled him greatly. As far as he could see, by "the scientific method" educators meant "something far more general than the methods by which the natural sciences have advanced: it is proclaimed as a way of looking at life; at times it seems almost a panacea for social problems."[28]

His assessment of the situation was spot on. The course he envisioned had to get closer to the work of the scientists themselves, and—contributing his piece to the NSF debate—he didn't mean to include those in the so-called social sciences. "I believe no good educational purpose is served by spreading the ambiguous title 'social science' over all the academic activities concerned with a study of man," he asserted. "And positive harm is done by claiming that the scientific method is going to save us. Indeed, something close to fraud is being perpetrated when this method is defined by implication as the process by which the physical and biological sciences have reached their present stage."[29]

The plan Conant came up with was to use a series of cases from the history of science (borrowing the case study method from the business school) to illustrate the way scientists pursued their work. His intent was for students to "study examples of science in the making." Given the complexity and level of theoretical sophistication that science had achieved to that point, he felt it would be difficult for students to draw on current science for their examples. Using science from the past had the advantage of providing instances that were simpler in their conceptual and mathematical content and, in addition, allowed students to see "in clearest light the necessary fumblings of even intellectual giants when they are also pioneers." "One comes to understand what science is," he added, "by seeing how difficult it is in fact to carry out glib scientific precepts." Among the historical cases were Robert Boyle's seventeenth-century work on pneumatics, the discovery of oxygen by Antoine Lavoisier and Joseph Priestly in the eighteenth century, and work related to the caloric theory of heat.[30]

The goal was for students to come away not with some view of scientific method but rather with an understanding of what he called (drawing on the military terminology prevalent then) "the tactics and strategy of science." Students would learn such things as the importance of new techniques in science and how they can revolutionize a field, see the complex interplay between experiment and the development of scientific concepts, understand the role of controlled experiments, and appreciate even the interactions between science and society. There would be nothing on the so-called scientific method, and to hammer home this particular point, Conant went out of his way to lay bare the inadequacy of Pearson's *Grammar of Science* version of it. He took issue in particular with what Pearson insisted was the first step in the scientific process, which entailed the "careful and accurate classification of facts and observation of their correlation and sequence." It was a terrible oversimplification from which he dissented entirely. "It seems to me, indeed," he wrote, "that one who had little or no direct experience with scientific investigations might be completely misled as to the nature of the scientific method by studying this famous book."[31]

An experimental version of the new course, The Growth of the Experimental Sciences, was taught by Conant in the fall of 1947. It was offered the following year as Natural Sciences 4 (Nat. Sci. 4) with the help of colleagues from the chemistry and physics departments. Nat. Sci. 4 was one of five courses developed as part of the Redbook-inspired general education program in science at Harvard. The others were a course on science in contemporary American life (Nat. Sci. 1) led by Philippe Le Corbeiller, one that focused on scientific problems with a physical-science emphasis (Nat. Sci. 2) taught by the physicists Edwin Kemble and Gerald Holton, another that followed a more traditional history of science approach (Nat. Sci. 3) taught by I. B. Cohen, and one with a biological science emphasis (Nat. Sci. 5). The development and teaching of those courses launched the careers of some of the most influential scholars in the field, none more significant than that of Thomas Kuhn, who while working as a graduate teaching assistant for Nat. Sci. 4 alongside Conant developed the ideas that led to the publication in 1962 of *The Structure of Scientific Revolutions,* one of the landmark books of the twentieth century.[32]

Ironically, when it came to laboratory work, the general education approach to science teaching in these courses (with their emphasis on illustrating the tactics and strategies of science) continued the trend started by the progressive educators. There were large-class demonstrations conducted to illustrate important empirical findings related to the development of some scientific concept or another. (In one unfortunate incident in 1954, the *Harvard Crimson* wryly reported that Leonard Nash, a colleague of Conant's, "actively disproved the Phlogiston theory [of combustion] in Natural Sciences 4 yesterday morning by an unintentional explosion which slightly injured him, deafened his students, and shattered glass as far as the back row.") But increasingly, individual student lab exercises were left out of the curriculum. One of the biology professors at Harvard a few years earlier had wondered aloud about this trend, noting that many instructors, even those professing "that first-hand observation and experiment . . . are essential in any serious attempt to introduce students to the methods and scope of science," after a few tries ultimately ended up doubting its necessity. "Many of the physical scientists" on campus, he went on to observe, "seem to have abandoned the laboratory for students." The days of Edwin Hall setting the standard for school laboratory work across the country more than six decades earlier seemed only a distant memory to these Harvard science instructors of the early 1950s.[33]

●   ●   ●

Conant's science education project occupied much of his time during these years apart from his duties as university president. Not only was he committed to creating case-study materials and teaching his Nat. Sci. 4 course for undergraduates, but he also worked to spread his general-education approach in science to a larger network of scholars, professors, and high school teachers. With additional funding from the Carnegie Corporation, through the 1940s and early 1950s he organized a series of conferences that were devoted to the topic and brought together many of the leading young scholars in the history, philosophy, and sociology of science along with historically and philosophically minded science faculty. These individuals formed the nucleus of what would later emerge in the United States as the field of science studies.[34]

The conference papers presented at the Harvard summer school in 1950 were published in a volume edited by two Harvard faculty members, historian of science I. B. Cohen and science educator Fletcher Watson. It spanned topics including the essential ideas in the history of science, the role of philosophy in general education science teaching, and teaching for citizenship in a technical civilization. Given the guiding hand of Conant in all this, it wasn't surprising that the "scientific method" came in for heavy abuse. Nearly every paper devoted a paragraph or more to debunking the myth of a stepwise process. "There is no 'scientific method,' no simple, clear-cut, easily definable procedure which, once specified, may be followed as rote," Cohen stated plainly in his contribution to the volume. "That there is no royal road to scientific discovery may be seen in the multitude of baffling problems which the scientists of our own day have not solved and concerning which no one knows with certainty how, or in which direction, to proceed."[35]

This message was extended to a new audience in the summer of 1952: high school science teachers. At the core of that summer program was a course, Methods of Science, that utilized the case-history approach; the first three weeks of the course were taught by Conant himself. "Quite frankly," he wrote to a colleague, "I am doing this in order to spread 'my poison' as widely as I can among secondary school teachers who I am more and more convinced shape the thinking of a large share of the coming generation." He led off the course by asking the fifty-two teachers in attendance to write a paper answering the question "Is it useful to talk about the scientific method?" and, following their immersion in the case material over the weeks that followed, ended with a final-exam question that prompted them to write what they would say at a science teachers' conference if asked to state publicly "the essence of the methods of science." Conant deemed the experience a great success. (Presumably the teachers came around to seeing the process of science in terms of "tactics and strategy" rather than as some routinized method.) In a personal follow-up on the course to Charles Dollard, his friend at the Carnegie Corporation, he passed on what he learned: "I had my opinion confirmed that many of the teachers of science at the school level have an oversimplified idea of the nature of the physical and biological sciences." And he seemed to have a pretty good idea where it came from: "This reflects what I believe to be a distortion of the John Dewey point of view."[36]

In addition to his local teaching initiatives, Conant kept up his drumbeat of public criticism of the scientific method during these years. He followed his publication of *On Understanding Science* with what was essentially a revised and expanded version of that work, *Science and Common Sense*, which was published in 1951. A year later came *Modern Science and Modern Man*, a book based on his Bampton Lectures that dealt with issues concerning the relationship between the public and science policy in democratic societies. Naturally it contained his well-rehearsed debunking of the scientific method. In *Science and Common Sense* he devoted an entire chapter to what he called the "alleged scientific method," and he even went so far as to solicit Dewey's thoughts on his critique, sending him a copy of the book to look over.[37]

Figure 14. James B. Conant on the cover of *Science and Common Sense* (New Haven, CT: Yale University Press, 1964). Reproduced from a copy in the author's collection.

Conant never actually faulted Dewey for the prevalence of the five-step method. His note to Dollard shows he knew he was dealing with a bastard-ization of Dewey's ideas. Even Dewey himself tried to correct the view that reflective thought could result from following a set of steps in some rigid fashion. In his revised version of *How We Think* published in 1933, he changed "steps" to "phases" and added a new section with the explicit heading "The Sequence of the Five Phases Is Not Fixed." But the idea had been out in the world for too long, and the potential for misattribution by others was ever present. William Kent, a philosopher from Utah, wrote to Conant upon reading *Science and Common Sense,* commenting, "In your excellent argu-ments against a unified method of inquiry, I was glad that you picked on Karl Pearson. But I was afraid that you had John Dewey in mind, since he has written so much on this subject." Conant had, in fact, read a good deal of Dewey's work and admitted his intellectual debt to the philosopher. His only regret, he replied to Kent, was that Dewey had "placed so much em-phasis on the steps of inquiry" and "failed to do justice to the interplay between conceptual schemes and experimental observations." For his part, Conant did what he could to set the public straight on this long-held misunderstanding.[38]

Conant found a sympathetic audience among his many readers (all three books went through multiple printings). His critique seemed to tap a reservoir of latent hostility to the commonplace view of the scientific method in particular. Philip Morrison, the Cornell University physicist, noted with satisfaction in his review of *On Understanding Science* that Conant "quietly but firmly destroys the facile position of those who maintain that science . . . is chiefly a set of simple logical recipes for the careful classification and ordering of facts." Another reviewer, a philosopher, praised *Science and Common Sense* for "briefing the busy citizen on the way in which science *really* works . . . [e]ffectively demolishing conventional twaddle about *the* scientific method." A reviewer writing for the *Quarterly Review of Biology* concurred, suggesting that this was indeed the book's "most stellar achievement." And Vannevar Bush—not surprisingly, given his own views on the matter—praised him in the pages of the *Saturday Review* for making it "crystal clear that there is no such thing as the scientific method." "The elegant definition of *the* scientific method that we have read for years," he went on, "comes in for the dissection it has long needed."[39]

The general education approach pioneered by Conant—particularly his emphasis on using the history of science to illustrate the circuitous path scientists often took in their work—had limited immediate effect at the high school level. The teachers who attended the 1952 summer school course may very well have carried those ideas back to their classrooms to begin shaping the thinking of the coming generation. And in the late 1950s and 1960s there was a short-lived effort to replicate Conant's method in high schools through the development of simplified history-of-science cases designed to supplement high school science teaching. But beyond this, there was little significant impact, at least none that can be easily traced. However, Conant's effort did mark a key turning point in the history of school science education in the United States: the return of the active involvement of practicing scientists in high school affairs following nearly three decades of control by progressive science educators. Following Conant's lead, research scientists in nearly every discipline turned their attention to the schools during the 1950s. Their goal was nothing less than a thorough reconstruction of curriculum and teaching to correct the misrepresentations of the past and to present what they saw as the true nature of scientific work.[40]

7 Origins of Inquiry

ON THE MORNING of October 5, 1957, the United States woke up to the news that the Soviet Union had successfully launched the world's first artificial satellite. "REDS FIRE 'MOON' INTO SKY," read the all-caps, front-page headline of the *Chicago Daily Tribune*. The satellite, called Sputnik by the Russians, reportedly circled the earth once every hour and thirty-five minutes, 560 miles up in space, and emitted a regular series of beeps that were easily picked up by ham radio operators across the country. The National Broadcasting Company interrupted its regularly scheduled programs to air the transmissions. "In thus announcing the launching of the first earth satellite ever put into globe-girdling orbit under man's controls," reported the *Los Angeles Times,* "the Soviet Union claimed a victory over the United States." It was a defeat that Americans felt to their bones, one that hit hard the belief that the United States was the world leader in science and technology.[1]

The years leading up to the fall 1957 launch had been tense. There was the growing scientific and technical challenge posed by the Soviet Union as the cold war began heating up, a challenge made more urgent by the spread of international communism and the Soviets' ability to keep pace with American nuclear weapons development. At home, worries over atomic espionage and charges of subversion among politicians, scientists, and teachers unsettled the public and were exacerbated by the reckless accusations of Wisconsin senator Joseph McCarthy, the iconic Red scare bogeyman of the early 1950s. In addition, though, the Sputnik launch, calling into question American intellectual superiority as it did, opened up an entirely new field of national concern—the public schools.[2]

Since the late 1940s, public education had been subject to a repeated pattern of low-level critique by various activists on the right. School over-crowding and the need it generated for increased tax revenues to fund district building projects garnered plenty of negative attention. Other groups targeted schools as dens of communist activity. Still other, less politically radical critics took issue with the dominance of the professional education establishment, as it seemed to control all aspects of the educational process in the interests of social adjustment, to the detriment of traditional learning. This was the critique that seemed to most capture the public mood at the time. Sputnik threw a harsh spotlight on what many viewed as the soft cur-riculum built upon the psychological pillars of progressive education. The emphasis in the schools on the students' life problems—their ability to fit in with others and feel good about themselves in 1950s America—suddenly

Figure 15. Detail from front page of October 5, 1957, *Los Angeles Times* with headline about Sputnik: "Russia Launches First Earth Satellite 560 Miles into Sky." Granger Historical Picture Archive / Image 0131594.

seemed to local and national leaders (if not all of the students' parents) dangerously naive in a world where Soviet high school students were mastering calculus, physics, and foreign languages.[3]

Pressure was felt everywhere, from community centers to newspaper editorial offices to the nation's capital. When Congress convened in January 1958—"SPUTNIK CONGRESS OPENS!" proclaimed the Chicago paper on its front page—it was clear that science education would soon be entering a new phase. "Russia's science education has dramatized as perhaps nothing else could the gap between our country's education needs and its educational effort," said an Evanston principal invited by the paper to comment on the current state of affairs. A local college president offered his views, hoping that "citizens and the Congress will give long and serious consideration to any course of action which represents an abrupt departure from long established and proven practices. . . . This applies especially in the field of education," he noted. "Abrupt" was the right word. By 1960 the federal government was funding no fewer than four large-scale curriculum reform projects in high school science with the direct involvement of at least five current or future Nobel Prize–winning scientists.[4]

The primary task of these projects was to produce curriculum materials that would give students a deeper understanding of the process of science as research scientists saw it. While nearly all the participants could agree that teaching the "scientific method" of the past was the last thing they wished to do, there was less agreement about what might replace it. The scientists knew that students needed to get away from everyday problems and back to what they believed to be "real science," to engage with natural phenomena in the laboratory or by some other means. But the materials they developed, at least initially, weren't guided by any well-defined notion of the scientific process. Most seemed content with Bridgman's notion of "doing one's damnedest with one's mind no holds barred."[5] By the early 1960s, however, a view of science did appear—the inquiry method. It emerged from the experiences and work of two key figures actively developing the new, post-Sputnik biology curriculum, Johns Hopkins geneticist Bentley Glass and University of Chicago professor Joseph Schwab. Together they helped advance a vision of science that would serve as the pedagogical cornerstone of science education, in theory if not fully in practice, through the 1990s.

●   ●   ●

As much as Sputnik provided the spark for wide-scale science education reform in the United States, it was, in fact, neither the initial nor the ultimate driver of change, as many have commonly believed.[6] Curricular reforms were already well under way in the physical and biological sciences when the public woke up to the news of the Russian satellite that October morning. The initial impetus for a new approach came from the situation in which research scientists found themselves in the federal patronage system after the war—the same circumstances that had motivated Conant and others in their work in science education at the college level. The effect of Sputnik was to excite public efforts in terms of both political support and material support (that is, federal funding) to remake high school science education in this new age of science.

The first of the reforms came in physics and was led by the MIT physicist Jerrold Zacharias. For Zacharias, the opportunity to get involved in the schools presented itself during meetings he attended as a member of the Science Advisory Committee of the Office of Defense Mobilization, a group established in 1950 that was responsible for national security matters at the highest levels of government. As Zacharias told it, in those early years of the cold war "the military would come in and complain that the Russians were getting ahead of us, that we had to do something about education . . . getting more engineers, more scientists." With personal entreaties from the leadership at the National Science Foundation and the promise of nearly unlimited funding, Zacharias launched the Physical Science Study Committee (PSSC) in 1956. His aim was to develop an integrated system of physics instruction that would include a textbook, instructional films, laboratory exercises, and a range of short books on current topics for student enrichment.[7]

Around the same time, researchers in the biological sciences began exploring ways to improve high school teaching in their subject in a series of modest initiatives. In 1953 the National Academy of Science's (NAS) Division of Biology and Agriculture sponsored a group to begin looking into new curricular materials that accurately captured modern biology in the postwar era. Soon thereafter the NAS established the Committee on Educational Policies to begin more substantive work along these lines, with funding from the Rockefeller Foundation. The most significant product of these efforts was the development of a sourcebook of laboratory and field study material prepared at Michigan State University in the summer of 1957 with the help of high school and college biology teachers. That same summer a group of college and high school chemistry teachers met at Reed College in Oregon to explore new approaches to chemical education. But the biology and chemistry initiatives did not have the level of inertia of the physics project, and apart from these, there were only scattered attempts to attack the problem of science education in any concerted way. The comparatively slow pace of this work, though, ended with the launch of Sputnik.[8]

Coming on the heels of a 1955 report on Soviet professional manpower that showed heavy Russian emphasis on high school science teaching, Sputnik galvanized legislative leaders to seek increased funding for education. In 1958 Congress passed the National Defense Education Act (NDEA), which aimed to improve the training of scientists and engineers in the United States. The science education budget at NSF was boosted to unprecedented levels, going from a baseline of just over $2 million in 1955 to nearly $80 million by 1962—a remarkable, almost fortyfold increase. The start NSF had made in physics with Zacharias's PSSC project quickly expanded to include all the high school science subjects. By 1964, well over $16 million had been spent on revamping curriculum materials in biology, chemistry, and physics—the core high school science subjects. Along with PSSC, NSF funded the Biological Sciences Curriculum Study (BSCS), led by Bentley Glass and Florida zoologist Arnold Grobman, and two projects in chemistry, the Chemical Bond Approach Project (which began with the Reed summer conference), headed by Laurence Strong at Earlham College, and the CHEM Study project, chaired by Nobel laureate Glenn Seaborg and directed by J. Arthur Campbell at Harvey Mudd College. The federal patronage system for research science now fully enveloped classroom science education as well.[9]

While newspaper and magazine editors, members of Congress, and other public figures clamored for schools to produce more scientists to meet the Soviet threat, the scientists who had stepped up to remake the curriculum were more interested in promoting greater public understanding of what science was. Their target was the students who would ultimately hold the reins of power over scientists and the scientific enterprise in their future role as United States citizens.[10] In this they followed the lead of Conant, who had sought the same outcome with Harvard undergraduates a decade earlier.

A primary goal of the PSSC course, for example, was to demonstrate for students how scientists think. Zacharias believed this was essential for the proper functioning of democracy and, moreover, would help combat the anti-intellectual hysteria of the time. The textbook and course materials PSSC developed focused on the epistemological question of how scientists know what they know; as he explained, it was all about "observation, evidence, and basis for belief," for this was "basic to everything." "Having lived through World War II, Hitler, Stalin, Joe McCarthy . . . the American public being molded by Joe McCarthy," Zacharias recalled, it was perfectly clear "that you have to understand why you believe what you believe." There was, of course, no well-defined method to be learned. The existing textbooks Zacharias and his colleagues surveyed at the time were, as they saw it, full of nothing but "false ideas" about the methods of science. Most of the physicists on the project were happy to drop method talk from their textbook entirely, keeping Bridgman's account as a thumbnail guide to everything they did.[11]

The biologists were similarly interested in conveying a more authentic picture of research in their field. This was especially important given the postwar transformation of work in the life sciences from the description and classification of specimens characteristic of the natural history methods of the past to more cutting-edge experimental work. The biologists, in line with most scientists then, wanted to avoid any association with the increasingly discredited "scientific method" that had long defined science in schools. A National Academy of Sciences official in charge of evaluating existing biology curricula found their treatments of the process of science exasperating. "Is it asking too much," he complained to a colleague, "to give individuals a real sense of the power of these methods of inquiry, and of their value and importance[?]" Textbooks, he insisted, needed to pay more attention to the relationship between content knowledge and process. "Introductory chapters dealing with the scope of a subject, its relation to others, and 'scientific method' would perhaps better be omitted, in many cases, so little do they show any real effort to find a conceptual framework . . . or appreciation of what is involved in scientific investigation beyond 'the method' set forth in Deweyite simplicity." Marston Bates, the Michigan biologist who would become the primary author of one of the BSCS textbooks, was equally dissatisfied with the way textbook authors typically described the work of scientists. "I wonder," he wrote to a friend, "if the fellows who teach biology in our country really believe the crap about 'scientific method' with which they uniformly start their textbooks."[12] It seemed that few of them did.

As these reformers set out to remake American science education during these tumultuous years, they did so with a keen eye on the public perception of their work. The first step was to undo what decades of formal schooling had done in drilling into students the idea that science operated by some step-by-step procedure known as "the scientific method." There were some indications that the new message was getting through. In its fifty-ninth yearbook, titled *Rethinking Science Education,* the National Society for the Study of Education (an organization that had long represented the views of the educational establishment) made only a single reference to the scientific method, and that reference denied its existence. "In the present century scientific methods and attitudes have been emphasized, particularly by Dewey, as objectives of formal education," wrote one of the contributors, and "there developed through the years an interpretation that there was one scientific method with definite steps to be followed in a sequential order. . . . Conant and others," he added, "have pointed out that this is not an acceptable interpretation."[13] What was acceptable, though, had yet to be fully determined.

•   •   •

As the curriculum projects got under way, among the very first activities was the creation of laboratory exercises, experiments, and films that would teach students about the nature of scientific thinking in each discipline. Most of the scientists involved were guided by only a tacit framework of just what it was about the process of science they hoped to convey, relying more on their own experience and instincts about what was important. BSCS, the biology project headed by Glass and Grobman, was the exception. Members of the Steering Committee spent considerable time discussing the specific learning outcomes they aimed to achieve. At their first meeting, held in February 1959 in Boulder, Colorado, the biologists settled on a set of general themes that would run throughout all the materials they planned to develop. The first one on the list was the "nature of scientific inquiry."[14]

The use of the word "inquiry" to label the process of science appeared to emerge from a particular set of historical circumstances. The interference in their work that scientists had worried about after the war and during the negotiations leading up to the establishment of the NSF were realized in more damaging and demoralizing ways during the Red scare of the late 1940s and 1950s, as scientists were repeatedly harassed by those suspicious of their allegiance to the United States. Worries that scientists were either deliberately working with the Russians or might inadvertently let atomic secrets slip in the course of their various intellectual exchanges with other scientists led national security officials to restrict scientific communication and intensify their surveillance. Scientists were hauled in front of the House Un-American Activities Committee, were subjected to loyalty tests, were refused passports for international travel, and had their telephones tapped and their letters intercepted. "There seems to be a well-organized campaign to paralyze all independent thought, discussion, dissent and protest in America, and men of science are conspicuous among the targets," raged AAAS president Kirtley Mather at the association's meeting in 1952. The government's treatment in 1954 of the brilliant physicist Robert Oppenheimer was particularly galling to scientists of the time. "The witch-hunting inspired by Senator McCarthy has sunk to depths heretofore considered out of reach of mortal man," one newspaper editorialized.[15]

As the scientists individually and through their various professional associations rose to defend themselves—to beat back "the attacks of those self-appointed watch-dogs of patriotism now abroad in the land"—they used the language of free inquiry. In an address celebrating the seventy-fifth anniversary of Ohio State University, MIT president Karl Compton sought to educate his listeners about the dangers of secrecy: "Science flourishes and progresses in an atmosphere of free inquiry and free interchange of ideas. . . . Any element of secrecy which is imposed on this activity acts like a brake to progress." Others followed. Detlev Bronk, the president of the National Academy of Sciences, deplored government restrictions on science and called for the country to "champion the freedom of scientific inquiry." Similarly, the *Los Angeles Times* reported that Caltech president Lee DuBridge "protested any fettering of intellectual or scientific inquiry" in a commencement address he gave in 1950. And the president of Columbia University insisted that "the fullest freedom of inquiry is still the strongest shield of a free and democratic people." Even Conant adopted the language of inquiry when he predicted optimistically in his 1952 Bampton Lectures that the "public would be able to comprehend the basic methods of scientific inquiry in the not-too-distant future." (Perhaps he was thinking of the new general education courses in science he was immersed in at the time.)[16]

These individual pronouncements were backed by the full weight of the executive committee of the AAAS, which sought to shift the focus of the association from the advancement of specialized research to actively managing

FREEDOM OF INQUIRY FOR SCIENCE URGED

Dr. Bronk of Johns Hopkins University Denounces the 'Trend Toward Secrecy'

CALLS FOR A UNITED FIGHT

Academy of Sciences Dedicates Its New Headquarters in the Former Woolworth Home

Dr. Detlev W. Bronk, president of Johns Hopkins University, Bal-

Figure 16. Detail from article titled "Freedom of Inquiry for Science Urged" newspaper headline from *New York Times,* April 12, 1950, 29. Reproduced from a copy in the University of Wisconsin Libraries.

the relationship between science and society. "In our modern society," the committee insisted, "it is absolutely essential that science—the results of science, the nature and importance of basic research, the methods of science, the spirit of science—be better understood by government officials, by businessmen, and indeed by all the people." Soon thereafter, scientists convened in Hamburg, Germany, to explore the topic of science and freedom, seeking through their public discussions "to make Western scientists and scholars more fully aware of what is entailed in the claim to freedom for the pursuit of truth." And in 1954 the American Psychological Association established the Committee on the Freedom of Enquiry and issued a report on the current state of the matter the following year.[17]

Equating scientific inquiry with intellectual freedom proved to be rhetorically persuasive in the popular discourse of the time. The idea of freedom of inquiry enabled scientists to push back against state control and, at the same time, shielded them from being labeled as Reds. Scientific leaders worked hard to impress upon the public, government funding agencies, and the national security apparatus that science needed to be free from restriction and interference if knowledge was to advance as rapidly as possible—and thus safeguard the nation from international military threats. As for concerns about communist subversion, they argued that science was incompatible with ideological dogma. Communism was well known for placing party loyalty above truth. Science, however, at its very essence was concerned with only the truth. "It is this freedom to think, to pursue lines of inquiry and draw objective conclusions regardless of social dogmas," explained *New York Times* science writer Waldemar Kaempffert, "that chiefly distinguishes the Western attitude toward science." The Soviet intellectual, on the other hand, was bound by a commitment to a political objective, which "makes it impossible for him to exercise the free criticism he would engage in were he loyal to the principles of scientific inquiry." Good scientists simply couldn't be communists.[18]

More than any biologist of his day, Bentley Glass was acutely in touch with the dangers threatening American science. From the late 1940s through the 1950s, he served on the editorial board of the AAAS, which oversaw the publication of both *Scientific Monthly* and *Science*—journals that regularly reported on the political attacks on science. In the summer of 1955, Glass chaired a special committee of the American Association of University Professors charged with reviewing dismissals of university faculty on the grounds of communist subversion. Reporting on the committee's findings, he affirmed the absolute commitment society must have to academic freedom: "Without freedom to explore, to criticize existing institutions, to exchange ideas, and to advocate solutions to human problems, faculty members and students cannot perform their work, cannot maintain their self-respect. Society suffers accordingly."[19]

Not only had Glass seen up close the pernicious effects Red-baiting had on scholars and scientists, but he also demonstrated a sophisticated understanding of the impact government (especially military) funding had on scientific work. As worrisome was the danger posed by ideological tests, such as those that had resulted in the rise of Lysenkoism in the Soviet Union. In the summer of 1957, before the general meeting of the American Institute of the Biological Sciences—just a year before the launch of BSCS—Glass called on biologists to defend their livelihood from all these impending threats. "From its earliest beginnings the inveterate foe of scientific inquiry has been authority—the authority of tradition, of religion, or of the state," he proclaimed. "No door must be barred to its inquiries, except by reason of its own limitations."[20]

This commitment to inquiry in the broadest sense was central to Glass's vision for the BSCS project, a vision shared by nearly all the members of the group. In his initial comments to the Steering Committee at their first meeting in 1959, Glass explained that "our job . . . is to look at the teaching of biology in the elementary and secondary schools to see how we can modify it. We should educate Americans, in general, to the acquisition of a scientific point of view." In a proposal they drafted for the National Science Foundation, this emphasis on promoting understanding of scientific thinking was unambiguous: "For most American secondary school students biology is the only science course taken. Thus it is especially important that through biology the general student learns to appreciate the growth of scientific knowledge and acquires a conception of the basis of scientific thought."[21]

While the biologists debated nearly every aspect of what their approach to high school biology would be, the focus on inquiry was constant throughout. At the third Steering Committee meeting, in New Orleans, Glass reiterated "that there were two emphases which should be very strongly developed: the nature of science as inquiry and as a growing self-correcting historical body of knowledge." What this meant, precisely, wasn't always clear. Columbia University zoologist John Moore took a stab at a more formal definition to share with the group. It included knowing, he wrote, "that science is an open-ended (ever expanding) intellectual activity and what is presently 'known' or believed is subject to 'change without notice'; that the scientist in his work strives to be honest, exact, and (part of a community) devoted to the pursuit of truth; that his methods are increasingly exact and the procedures themselves are increasingly self-correcting."[22] It was a start.

•   •   •

Joseph Schwab joined BSCS in November 1959.[23] As a member of the faculty of the natural sciences and education at the University of Chicago, Schwab came to BSCS with an impressive portfolio of innovation in science education. Although Glass and other members of the steering committee had gotten the ball rolling, Schwab quickly exerted his influence on the project. His intellectual background and dedication both to a deep understanding of the workings of science and to education for public understanding of science resulted in the crystallization of the group's ideas and language about inquiry and its integration throughout all components of the project mate-

Figure 17. Joseph J. Schwab, 1952.
University of Chicago Photographic
Archive, apf1-07509, Special
Collections Research Center,
University of Chicago Library.

rials. Schwab, more than anyone else in the twentieth century, defined for a generation what scientific inquiry was and what teaching science as inquiry essentially entailed.

Schwab spent nearly his entire academic life at the University of Chicago. He enrolled as an undergraduate at the age of fifteen in 1924, initially pursuing a major in physics. But after struggling with the required mathematics and the rote methods common in physics at the time, he turned toward literature and eventually graduated in 1930 with a bachelor's degree in English. He subsequently took up graduate work in zoology, earning a master's degree in 1936, and then spent a year on fellowship at Teachers College in New York before returning to Chicago to teach in the biological sciences sequence of the general education program while completing his PhD in genetics under Sewall Wright. In 1941 Schwab took a position as assistant professor of natural sciences, and soon after that, as a result of his pedagogical work in the general education program, he received an appointment in the education school as well (though he lived more as a scientist / philosopher who was interested in general education than as someone closely identified with the professional education establishment). During these years he embraced the challenge of general education and became a leading figure in the natural sciences program at the university and nationally.[24]

Almost from the beginning of his work with the biology sequence at Chicago, Schwab expressed frustration with the haphazard way in which introductory science courses were typically taught. Science faculty, he felt, were interested only in reproducing the technical aspects of their disciplines in whatever way would detract least from their research. Schwab, however, considered the objectives of such courses from the standpoint of the learner, and in doing so concluded—much as the progressive science educators had before him—that what was needed was not the technical knowledge of specialized science but rather knowledge of its methods. "This then is one place of the sciences in the education of men and citizens," he explained in a 1941 essay on the objectives of natural science instruction. "It is not the content of sciences, but their natures, that the citizen must know—what they are good for and what their methods are." It was apparent to Schwab that there was far too much science content than could ever be covered in any useful way. However, there was practical value in understanding the methods by which disciplinary knowledge came to be.[25]

In 1942 Schwab signed on to a committee tasked with recasting the introductory science courses at Chicago as part of an effort by university president Robert Hutchins to reform undergraduate teaching. As recounted in a 1948 essay describing the program, the committee designed the new courses to focus on the process of science, much as the existing general education courses in science had done. But rather than being organized around specific disciplines (either biological or physical science), the new courses were part of a three-year program in the "natural sciences." When the committee members sat down during their planning meetings to share how each of them thought about the scientific method and had typically approached teaching it, they were startled to find that they "differed to an extraordinary degree on what constituted scientific method itself."[26]

This led the group to an intensive, months-long study of writings by philosophers and scientists on the nature of science, at the conclusion of which the members arrived at some key insights. Of the numerous statements and doctrines about the scientific method they had found, many were merely vague generalizations or aphorisms characterizing science as "the attempt to seek the objective truth free of bias and emotion," and others were aimed at reducing science to a set of equally vague, predetermined steps (the "scientific method" of the progressive educators). Still others were found to be true only in specific instances, which was the case with Pearson's definition of

science as the "classification of facts," and failed to accurately reflect the whole
of scientific work. The most obvious conclusion of their study was that (as
many scientists of the time had been suggesting all along) "there is not one
scientific method, but several, differing from field to field, and even, in many
cases, from problem to problem."[27]

More important than their finding that no single method captured the
intellectual process of scientific research was Schwab's realization that any
consideration of method must necessarily entail consideration of the scientific content related to a given research question. That is, method and content were mutually dependent. As Schwab put it, "Content was not only
indispensable to *understanding* method, but that—shock to our semantical
habit—method *had* a content." He explained that "if we segregated all that
was not *process* under the rubric *content,* and ignored it for the purpose of
understanding method, a distorted or incomplete view emerged of how the
scientist 'found' or 'developed' knowledge." "There appeared to be a substantive element in method," he went on. There were "tools for the hand and
the senses, . . . the commonly recognized *instruments* of the laboratory," but
there were also "instruments employed by the mind as aids in its share of the
scientist's work"—the concepts and theories that determined which questions
could be asked and what might count as a possible answer.[28]

This point on the contextual nature of scientific inquiry—the committee's finding that it was impossible to have a free-floating method divorced
from disciplinary content—reflected, ironically, the influence of John Dewey
on Schwab's thinking at the time. As he became more engaged in thinking
about curriculum development in the 1940s as chairman of the natural sciences group, Schwab had begun reading Dewey and was taken with his views
on the importance of context in the pursuit of knowledge. Dewey had long
sought to dispel any view that might separate knowing from experience. In
his 1938 book *Logic: The Theory of Inquiry,* he emphasized that "since every
special case of knowledge is constituted as the outcome of some special inquiry, the conception of knowledge as such can only be a generalization of
the properties discovered to belong to conclusions which are outcomes of
inquiry." In other words, for Dewey, knowledge (or science content) was not
something that existed in isolation, separate from experience. Instead it was
the culmination of some inquiry that had value in its availability "as a resource in further inquiry." This notion of the situated nature of knowledge
appeared throughout Schwab's writing. It could easily have been Dewey

rather than Schwab who wrote that scientific words or concepts "take their most telling meanings . . . not from their sense in isolation, but from their context." The meaning of "free fall, of electron or neutrino, is understood properly only in the context of the enquiry that produced them."[29]

Schwab's affection for Deweyan thought, however, stopped short of the five-step scientific method that had infused high school science teaching. In attempting to lay out what he believed to be a useful way to think about inquiry, he began with the common misrepresentation of science. "Let us begin," he wrote in an essay on the nature of disciplinary knowledge, "by examining the description of science which has its origin in John Dewey's *How We Think.* . . . This starting point is almost mandatory, for the Deweyan formulation, which describes science as taking place in a sequence of steps," he went on, "has appeared and reappeared in so many textbook prefaces that it has taken on an official character." A single, step-by-step method of science simply did not exist, as he knew from his work on the curriculum committee at Chicago and his own experience first in physics as an undergraduate and then in genetics later on. He went on to offer a more sophisticated account of how scientists did their work in which the existing structures of disciplinary knowledge provided a framework for asking questions, seeking data, and finding answers.[30]

In providing this overview, he made a distinction between what he called "stable enquiry," which proceeded using the existing conceptual knowledge and principles of the discipline, and "fluid enquiry," which occurred when the concepts being used to pursue inquiry were no longer fruitful. In those instances, scientists proceeded to reexamine the foundational principles themselves. The very progress of science, according to Schwab, resulted from alternating periods of stable and fluid inquiry. The "schoolbook version of science," as Schwab called it, cut off from content, had no way to account for the role disciplinary knowledge played in providing the framework for the conduct of research, and its five steps were essentially useless when it came to describing the process of fluid inquiry, when all structure broke down and progress depended on the scientist's creative imagination and insight.[31] Make no mistake, this account of scientific work was thoroughly grounded in Dewey's work; what Schwab rejected was the simplified scientific method others had long attributed to him.

Schwab's ideas on inquiry provided a fitting complement to Glass's thinking about science teaching and the public. Indeed, the affinities are easily seen in articles Schwab and Glass each contributed to a special issue of the *Bulletin of the Atomic Scientists* on the theme of "Science and Education" in November 1958. In Schwab's piece, he reiterated the importance of students understanding scientific conclusions not as fixed truths (i.e., dogma) but as emerging from the "context of inquiry which structured and bounded the matters to which they refer." The focus on conclusions alone in science classrooms resulted in a "perversion of the meaning of science." Glass too lamented the emphasis on masses of facts. "Legions of our science textbooks," he wrote in his contribution, "serve up to hapless students a crystallized, anonymous science that seems to have descended perfect, like the divine city out of heaven, straight from unquestionable authority." The mindless inculcation of textbook content, he insisted, "is one thing that should be anathema in the sciences." It was evident to both Schwab and Glass that in the postwar world in which they lived nothing less than the future of science and democratic society itself depended on greater public understanding of the way science worked. Glass warned that without such understanding, "we face . . . eventual collapse of our form of civilization, our way of life."[32]

● ● ●

Although many scientists of the time were motivated to reshape science education during these years by social and political concerns, Schwab offered perhaps the most fully developed political argument for his approach to science teaching in his extended 1962 essay "The Teaching of Science as Enquiry." His worry about the precarious place of science was rooted specifically in what he came to understand as its fundamental character. It wasn't just that the mutually dependent relationship science had with the federal government had created a need to court public approval. It was that tremendous federal investment had greatly accelerated (and was likely to continue to accelerate) science's natural pace of advancement. Gone were the days when the most significant developments in science were "to be looked for in the sixth place of decimals," as Albert Michelson had remarked in his 1894 convocation address in Chicago. The years between Michelson's time and Schwab's had seen dramatic changes in the nature of knowledge and the speed at which it changed.[33]

In those earlier years, science could be taught as an accumulation of facts and concepts that would become only more precise with time—knowledge that was the outcome of Schwab's stable inquiry. But with the discovery of radioactivity, the development of quantum theory, and Einstein's relativity emerging during the first decades of the twentieth century, physics (if not science as a whole) had been turned upside down. Entire theoretical structures, once thought to be firmly established, were overthrown. Basic concepts of space and time were called into question. Science during these years underwent a remarkable period of fluid inquiry, in Schwab's parlance, and such periods, he felt, were going to come more frequently with the government-fueled expansion of the scientific enterprise. An "enormous acceleration of the revisionary rate in the sciences" was on the horizon, he thought, and with it came "the enhanced size and status of the fluid enterprise," which could no longer be "hidden." And this, Schwab believed, would present the public with an image of science that was at odds with existing beliefs about science as a steady, cumulative enterprise that over time moved closer and closer to the truth.[34]

In his essay, he explained how important it was that citizens not learn science as a form of irrevocable truth, or, as he famously stated, a "rhetoric of conclusions." In his mind, it was essential that a new type of science education be developed to teach about fluid inquiry. Citizens needed to "understand that it is a mode of investigation which rests on conceptual innovation, proceeds through uncertainty and failure, and eventuates in knowledge which is contingent, dubitable, and hard to come by." The need was urgent—"nothing before in the history of this country, short of war or disaster, has evoked such a concert of effort and attention." "Indeed," he continued, "factors closely akin to war and economic fatality are involved in the situation." At its essence, the goal was a political one—to educate the nation's high school students so that they might "maintain and support a mode of scientific enquiry which has never before been so urgently required, so visible to the naked, public eye, and understood so little by so few."[35]

Schwab envisioned a number of audiences for his vision of science teaching. There was a need for scientists to populate the ever-growing number of federally funded laboratories, individuals who could engage in the kind of creative, outside-the-box thinking that fluid inquiry entailed. There was also the need to educate future civic leaders and government officials who would be able to pursue public policy and make decisions regarding important socioscientific issues. These individuals, he felt, must recognize that scientific knowledge and expertise are flexible rather than rigid and dogmatic and that there may very well be differing expert opinions on any given issue. A political leader with the appropriate understanding of science would be able to negotiate among the various recommendations and help guide the decision process to an appropriate solution. (These were the same individuals—and goals—Conant had sought to address in his Nat. Sci. 4 course.) But the audience about which Schwab was most concerned was the lay public. Individuals in this category were vital to the democratic political system in that they held the power to control government action, and their education came primarily in the high school, where the new curriculum projects were targeted.[36]

Typical high school science teaching, according to Schwab, presented science in the "rhetoric of conclusions" vein, which "conveys the impression that the assertions of science are inalterable truths." But teaching science as a set of truths or facts—that electrons are real things, for example, or that enzymes always function in some specific way—sets up conditions for public confusion, even for a backlash against science, when the inevitable period of fluid inquiry occurs. What happens when those seemingly solid conceptual structures themselves are torn down and rebuilt? "Consider a student who has garnered the impression that science consists of inalterable truths," Schwab wrote. "Five or ten years after graduation he discovers that many of the matters taught him are no longer taught, nor any longer used as knowledge. They have become obsolete and been replaced by other formulations." He continued with his hypothetical story: "Unprepared for such a change, unaware of the operations of enquiry which produce it, the former student, now a voting member of the polity, can do no better than to doubt the soundness of his textbook and his teacher." Moreover, "in a great many cases, this doubt of teacher and textbook becomes a doubt of science itself." "The former student," Schwab submitted, "has no recourse but to fall into a dangerous relativism or cynicism," which, if spread among the public at large, would threaten the continued support and funding of the scientific research enterprise.[37]

There were a variety of pedagogical methods that could be used to help students see the nature of scientific inquiry as Schwab conceived it. His preferred approach was through reading and discussion, a method he perfected running the natural science courses in the undergraduate college at Chicago before joining BSCS. Indeed, when it came to laboratory work, those courses had a long history of using mostly demonstrations and exhibits. Students were, in fact, not allowed to participate in laboratory work or even handle apparatus. It wasn't that Schwab objected to lab work (and to be fair, there was some in the revised, three-year natural science program). Rather, reading and analysis, he felt, were more effective than individual laboratory experiences in helping students see the way ideas were developed and used in science.[38]

As for scientific inquiry, he made an important distinction between teaching science as a process of inquiry (the goal of which was to have students understand how science works) and teaching science by engaging in inquiry. The latter approach—using the process of science itself as its own pedagogical method, which was the approach that had justified the laboratory method of the late 1800s—had students doing science (in some form) as the means to learn science. As he wrote toward the end of his essay, "Of the two components—science as enquiry and the activity of enquiring—it is the former which should be given first priority as the objective of science teaching in the secondary school. It is a view of science as enquiry which is necessary if we are to develop the informed public which our national need urgently demands."[39]

Text- and discussion-based teaching methods, however, posed their own challenges. The most significant was the difficulty students with limited science backgrounds had making sense of the technical language in which the findings of scientific research were reported. Conant's solution to this, of course, was to reach into history and walk students through cases of scientific advancement from the past, counting on the fact that the "tactics and strategies" of science would be the same regardless of context. This was an approach Schwab knew well through his immersion in the general education movement and from having attended Conant's summer conference in 1949. (A member of the Harvard group listed Schwab in a follow-up report to the Carnegie Foundation as someone who "made a good impression" during the meeting.) In his undergraduate teaching at Chicago, however, Schwab elected to go a different route. For his instructional materials, he assembled a large number of original scientific papers for students to study. "Each of these papers," he explained, "infers, deduces, induces, defends, attacks, constructs, analyzes, proposes, or in some other way *moves* from '*data*' to '*conclusion*.'" He sought out examples—some from early science and some from contemporary research—that would illustrate, among other things, the plural nature of scientific methods, the variety of ways scientific problems can be attacked, and the current conceptual thinking about particular problems in any given discipline. His interest was in presenting students with actual "scientific papers as *instances of good enquiries yielding scientific knowledge worth possessing.*"[40]

Schwab's ideas for teaching about inquiry had a significant influence on the work of BSCS. Although the study of original papers didn't make the cut for the high school project, he was successful in his insistence that the textbooks the group developed present students with what he called a "narrative of inquiry" rather than a rhetoric of conclusions. Glass directed the writing teams to do all they could to highlight inquiry throughout. "Let us avoid presenting shovelfuls of authoritative facts and doses of scientific dogma," he wrote to the lead authors. "Rather, let us indicate clearly and frequently that scientific concepts and theories are based on *evidence*. Evidence springs from observations and experiments. . . . But inasmuch as error may enter into either observations and experiments or into the logical deductions based on them," he added, "scientific knowledge is necessarily imperfect and tentative."[41]

The results were promising. The group produced three different textbook versions of their high school course coded by color—yellow, blue, and green. Authors of the yellow version made inquiry a part of the title: *Biological Science: An Inquiry into Life*. And the blue version opened with four chapters that treated the interaction of facts and ideas in biology. The first of these, "Science as Inquiry," provided a detailed discussion of how scientists went about their work, using Darwin's theory of the formation of coral islands as an instance of scientific thinking in action. The green version, as well as the others, included scores of other historical and contemporary examples illustrating how scientific knowledge emerged and was actively built from a wide range of natural patterns, puzzling situations, and deliberate experiments. The books to varying degrees clearly displayed the "narratives" Schwab described in his writing on the topic. This was the sort of material that allowed students to "dissect the records of scientific inquiry in order to distinguish its constituent concepts, assumptions, data, etc., and understand their roles."[42]

Perhaps the most innovative pedagogical tool Schwab developed was his "Invitations to Inquiry." Drawing on his earlier efforts to catalog the diverse methods of science in the undergraduate curriculum at Chicago, Schwab developed a series of classroom exercises that fell somewhere between a straightforward textbook passage and a full-blown laboratory exploration. The "Invitations" were textual vignettes designed, in Schwab's words, "to *'show* the student that knowledge arises from data'; *'show* the student that knowledge changes.'" As he explained it, "The point is that the mere telling of students about such things is not effective. The need is to exhibit science in operation."

Each of the "Invitations" sketched a problem or situation, part of which was left unfinished for the student to complete after reasoned consideration of the material provided. As Schwab wrote, a given vignette taught inquiry in two ways: "First, it poses example after example of the process itself. Second, it *engages the participation* of the student in the process." The "Invitations" afforded students the opportunity to do the intellectual work without having to bother with the physical manipulation of either apparatus or living organisms. The students' responses would be followed up with teacher questions challenging and refining student ideas as the class was led to an understanding of the element of inquiry that was the focus of the scenario. In the realm of curriculum material, these exercises were unique. At the spring 1962 Steering Committee meeting, John Moore commented that the "Invitations to Inquiry" were "the most interesting things to come out of BSCS."[43]

Although using original papers was never seriously considered for the high school student audience of BSCS, and the "Invitations to Inquiry"— innovative as they were—were often passed over owing to their somewhat pedantic nature and the requirement for teachers to engage students in analytical discussions for which they weren't all that well prepared, Schwab's ideas about inquiry had a far-reaching impact on science education nonetheless. His insistence on the importance of presenting science as a process of knowledge generation, where the outcomes are closely tied to the context of an investigation that utilizes disciplinary knowledge as an intellectual tool, came to infuse the curriculum materials of the biology group. In this way, his ideas complemented the focus on inquiry as the essential feature of science espoused by Bentley Glass, the project chairman. Moreover, Schwab's extended writing about the nature of scientific inquiry more generally offered scientists and science educators an idea and a phrase that could effectively break the monopoly the "scientific method" had on the science education establishment, if not the public at large. "Inquiry" not only captured the complexities of scientific work but also aligned science with then-resonant political ideas of autonomy and freedom. Teaching science as inquiry would be the rallying cry for reformers going forward, lasting into the final decades of the twentieth century. Creating new learning experiences for students that conformed to this ideal in the short term was another matter.

8 Scientists in the Classroom

THE DISTINCTION SCHWAB made between teaching science as a process of inquiry and teaching science through the inquiry process was one that many reformers overlooked in the post-Sputnik rush to develop new curricula for American classrooms. Most of the scientists involved just assumed the two went hand in hand. If one was teaching students by having them engage in scientific reasoning through various inquiry activities, then it seemed a foregone conclusion that they would come to understand and appreciate the way in which science operated. While the BSCS group took care to think deliberately about the many ways science might be portrayed in the various parts of the curriculum, the curriculum projects in the physical sciences—such as PSSC, CHEM Study, and the Chemical Bond Approach Project—focused more on using laboratory experiences to convey ideas about scientific process. That is, they assumed that simply through immersion, students would come away with an understanding of scientific inquiry.

Formal laboratory study, neglected for the most part since the early 1920s, was a high priority of the new reformers, and its reintroduction into schools was their first order of business. As Bentley Glass explained in his characteristically lofty prose, "The laboratory is where the work of science is done, where its spirit lives within those who work there, where its methods are transmitted from one generation to the next." "No matter how much you learn about the facts of science," he went on, "you will never quite understand what makes science the force it is in human history, or the scientists the sorts of people they are, until you have shared with them such an experience."

The goal was the same for the physical scientists. The purpose of the new curricula, in the words of one physicist, was to have students "do physics the way *we* do physics." And this required an immersion in the ideas, tools, and materials of physics in the laboratory.[1]

The renewed interest in laboratory teaching received a tremendous boost with the passage of the National Defense Education Act in 1958, the signal legislative achievement of the "Sputnik Congress." Among its many provisions for strengthening America's schools was Title III, which authorized $300 million in federal matching funds in the first four years of the legislation for school districts to remodel and restock school laboratory facilities in science, mathematics, and foreign languages—subjects deemed essential to national defense. The investment, scientists felt, was long overdue. The infusion of cash, three-quarters of which went toward science projects in the first year, enabled schools to purchase equipment and teaching aids essential for implementing the new curricula being developed.[2]

The reemergence of the instructional laboratory was therefore a central feature of the new science education reforms. Local science teachers and science supervisors suddenly were awash in money for remodeling and new equipment. News stories of the time highlighted the radical changes that were being made with the new laboratory-based curricula. "Biological Sciences Curriculum Study Revolutionizes Lab Work," read the headline of a newspaper in Colorado. The laboratory experience was front and center in the physics project as well. A *New York Herald Tribune* article explained how the old-style, "cook book"–type labs were being replaced with "an imaginative laboratory workshop," where "paper, straw, old curtain rings, and almost anything cheap is used to make simple but highly successful experimental devices." A 1959 front-page article in the *New York Times* on PSSC similarly played up the innovative laboratory ideas from "a group of rebellious physicists and education specialists" who were "carrying the rebellion to more than 24,000 students in 600 high schools."[3]

The essence of the new projects was aptly captured by a reporter from the *Chicago Tribune* who wrote that "students tackle real scientific problems using the same methods scientists use."[4] That is, no longer were students to be taught the process of science so that they might be able to apply it in their daily lives. Instead, they were immersed in science—in the laboratory where scientists worked—so that they might better appreciate the complex material and intellectual work scientists engaged in as part of *their* daily lives.

Learning to "think like a scientist" was a necessary step toward understanding the nature of science itself as a self-contained professional activity. Numerous challenges, though, awaited the reformers. It was one thing to create opportunities for students to participate in scientific inquiry. It was something completely different to actually succeed in implementing that vision in classrooms and with teachers who had their own ideas about what the process of science was all about.

*   *   *

When the physicists and chemists got together in the mid-1950s and early 1960s to survey the state of teaching in the physical sciences, they were disappointed in what they found. Most textbooks, in their assessment, tended to cover too much material and spent too much time on technological applications and gadgets in an attempt to excite student interest. "Some texts," according to one report, "stress the technological aspects so strongly that they could be better labeled as textbooks of engineering than textbooks of physics." This, they felt, detracted from the coherence and unity of the subjects and pushed students (and teachers) into patterns of rote learning. Lost was a view of physics or chemistry as disciplines unto themselves. The physicists aimed for something much less applied. In their plan, "no emphasis would be placed on understanding the workings of modern technological devices as such," in hopes of presenting physics as "a product of human minds in the pursuit of truth, an activity which has evolved through the centuries and is still evolving and which has created a philosophical structure and a particular way of knowing." Their overarching goal was fostering a greater cultural appreciation of science and its methods, to teach science as a liberal art.[5]

The PSSC project opted for a multi-pronged approach to the task. They drafted a new textbook to lay out the essential ideas of modern physics, focusing in particular on an in-depth treatment of optics, waves, and mechanics along with the role of atoms in the physical universe. Paired with the textbook were a series of instructional films and laboratory exercises for the student. All three components of the course were designed to illustrate the way physicists engaged in research. Discussion of the scientific method was eliminated entirely. Instead, the "practice to be followed in the new course will be to make the basic concepts plausible by presenting selected evidence in the form of experimental demonstrations, films and textual material." They would avoid "attempts to follow rigidly any formal scheme of inductive reasoning." An "honest presentation of the methodology of science," they insisted, would need to include the "errors and excursions into unfruitful bypaths . . . that are often omitted." For Zacharias, a gifted experimental physicist, the process was recursive: "We do experiments from which we make a theory. The theory suggests new experiments. We perform the experiments. We modify the theory usually as the result of the experiments, and so on. I call it right foot, left foot, right foot, left foot," all the while gradually moving toward greater knowledge about the world.[6]

The films (approximately seventy in all) were used in tandem with the laboratory exercises to highlight this process. Each film dealt with some particular concept or experiment—typically something that couldn't easily be seen in the classroom, such as the pressure of light or the fundamental charge of the electron—and proceeded to demonstrate it for students, with real physicists on camera explaining how the evidence generated supported the theory or principle in question. The use of real scientists was deemed essential so that students could see them as normal people and not "as beards, hermits or quiz kids," as they put it. More important was the ability of scientists—through their actions on film—to show students (and teachers) how to reason with experimental evidence, to get the "logic lines" between the empirical data and the theory "crisp, clean, clear, and charming." In this way it would be possible, as Zacharias described in his initial memo outlining the project, to build the entire conceptual structure of physics from a continuous series of experiments.[7]

As important as the films were, it was the laboratory exercises that were key to the project. It was "the last thing in the world we would have given up," Zacharias remarked later. These weren't to be activities where students simply verified what was already known. The "experiments should indeed be experiments," the group insisted. According to the project's guidelines, "The laboratory should be open-ended in that each experiment should suggest and encourage further experiments along similar lines." In an attempt to put the responsibility for discovery on the students, Zacharias and his team provided as little laboratory direction as possible. "At the beginning of the course, some students may feel a little insecure with this type of laboratory work," they counseled the teachers. "They are likely to ask whether they have the right result. You must assure them that nature is not wrong; our job is to understand it by measurement and interpretation." The hope was that students would use their own creativity and initiative to wrest some truth from the phenomenon under study. "If physics is an open-ended, continuing process of inquiry, exploration, and discovery," they reasoned, "then the course should—as much as possible—be that way too."[8]

The physicists were committed as well to apparatus that was as simple as possible, better even if it was constructed by the students themselves using common household materials. This approach enabled them to foreground the intellectual process of science. "Contrary to popular belief," they told the students, "a laboratory need not be a sprawling glass-walled building bulging with giant electronic computers, huge cyclotrons, instrument panels, and row upon row of apparatus-laden benches. . . . Most times, the individual physicist bent on proving a point has to rely on equipment assembled from readily available odds and ends." Almost all the apparatus in the program was designed initially to be constructed by the students. This made the implementation of the course cost effective for school districts. But, more important in their eyes, it was an opportunity for the students to work as physicists did. A physicist, they explained, "improvises, he builds, he rebuilds, and he tries again until his equipment, whether it be simple or complicated, represents the best possible arrangement to provide the measurements and data he requires." Speaking directly to the student, they said, "You can improvise, you can build, you can invent, and you can judge the accuracy of your own doing, your own contriving." Their plan was to immerse students in the materials and phenomena. It wasn't about following some method; it was about doing their damnedest with their minds, as any good physicist did.[9]

The first draft of the laboratory guide led off with tutorials on how to cut, drill, and fasten sheet metal; how to cut and fasten plastics; proper techniques for bending glass and making pipettes and medicine droppers; and how to mix various glues and resins. Students were expected to use these skills to construct their own apparatus. They were told how to build micrometers out of boards, glass plates, and a sewing needle; make a stroboscope out of a vinyl record; and fabricate a highly sensitive microbalance from a soda straw, some matchbook covers, and rubber bands. But even these tasks were couched largely as suggestions, rather than as predetermined instructions to follow. *"Use the material in these laboratory books more as a guide than as hard-and-fast procedures,"* students were told. They were encouraged, above all, to play with the equipment, to explore how things worked. "When you have finished a suggested experiment always ask yourself this question: 'But

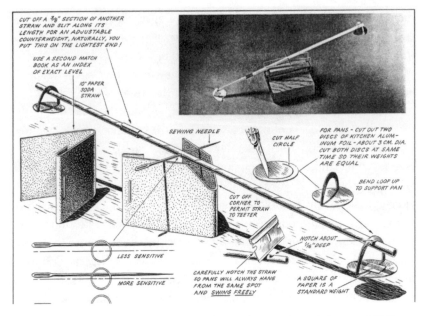

Figure 18. Diagram for the student construction of an equal-arm balance in PSSC laboratory program. From PSSC, *Physics, Laboratory Guide No. 1, Preliminary Edition* (Cambridge, MA: PSSC, 1958), 6-3. Jerrold Reinach Zacharias papers MC31: Box 4—Physics, Institute Archives and Special Collections, MIT Libraries, Cambridge, Massachusetts.

what would happen if I varied this or changed that?'" The goal for Zacharias and his team was to give students the space, materials, and opportunity to make sense of physics on their own. But this required students who were motivated to do so and teachers with the ability to guide students in the process.[10]

The initial trials of the course had mixed results. Many of the teachers reported that the laboratory work was "outstandingly successful" and that student enthusiasm was high. Some—typically the better students—even begged for more time to work on the exercises in class or took materials with them after class to continue the work at home. Most, however, struggled with the lack of guidance. In feedback sessions, teachers frequently asked for more "rigorous directions." Many found their students lost in the laboratory with no real understanding of what they were supposed to do. One teacher commented that students "need (but do not seem to have) [the] ability to 'dig-out' material on their own." As late as the winter of 1962, more than six years into the project, it seemed that "the problem of how to interest the student in the simple-minded playing with apparatus was . . . unsolved."[11]

Adding to the frustration of the teachers was the lack of clarity regarding laboratory write-ups. After decades of step-by-step laboratory work followed by the careful documentation of each step, students (and teachers) were at a loss when it came to recording student work in the PSSC laboratory. Teachers were clearly looking for more detailed expectations to keep students focused on their work, and despite the physicists' efforts to get students to just play and explore, teachers seemed to expect that science demanded some form of rigorous documentation.[12]

The requirement that students construct their own apparatus turned out to be particularly challenging. Both the students and teachers found the work time consuming and burdensome. Teachers reported that it was "needlessly involved and inaccurate." Some wondered "if the principles of physics involved in *construction* had sufficient value" when weighed against the time commitment. In response, one member of the PSSC team suggested that perhaps Saturday morning laboratory sessions might help with the time pressure, a suggestion that harked back to Charles Eliot's plans for teaching the Harvard physics course of the late 1800s. Other teachers simply "did not like the 'home-built' aspect of the laboratory equipment," worrying about things such as storage, the seemingly unscientific nature of it, and the ability of girls with "weak background[s] in this kind of work" to be successful. These concerns eventually led the PSSC lab group to create kits of materials to ease the process and, finally, to partner with some commercial apparatus manufacturers to provide a catalog of materials that schools could simply buy for their classrooms.[13]

The chemistry projects that came on the heels of PSSC were similarly interested in having the students experience natural phenomena firsthand in the laboratory. Existing textbooks, in the chemists' eyes, placed far too much emphasis on established concepts and theories, minimizing the "kind of direct laboratory work which would introduce the excitement of basic research and encourage motivation toward imaginative experimentation." As the CHEM Study project took shape, J. Arthur Campbell, the director, laid out the essential character of the course. It should build concepts from empirical evidence encountered by the student, and he insisted that the experiments that generated this evidence "be real experiments rather than demonstrations. The student should not, in general, know the answer to the experiment before it is begun, but should be able to make discoveries during the performance of the experiment." This would enable them to accomplish what they called "Goal No. 1" of the course: "To convey the experience of scientific activity"—their version of the scientific method—the essential components of which were "a. Observation and measurement; b. Inductive approach to general principles; c. Deductive use of principles in prediction; [and] d. The experience of discovery."[14]

The first chapter of the CHEM Study textbook presented what the authors called "the activities of science." These were listed in a series beginning with "observation and description," followed by a "search for regularities," and ending with "wondering why." Prior to classroom discussion of these activities or reading about them in the textbook, students were sent into the lab to investigate some chemical or physical phenomena, such as the burning of a candle or the melting of various solids, to serve as a foundation for understanding. (This led to students staring at a burning candle for an hour, much as Agassiz's students would gaze in solitude at a fish nearly a century before.) As it was explained to the teacher, the student "gathers some facts—that is, gets an experimental basis for understanding concepts—which the teacher then develops the next day." "This technique," they explained, "is superior to sending the student into the lab to verify generalizations made by the Textbook or the teacher."[15] Evidence that couldn't be gathered directly by the student was often provided through teacher demonstrations or in the accompanying set of films that were developed for the course.

The emphasis on the primacy of relatively accessible data early on set the stage for more sophisticated ideas that came later in the course. When a student is "asked to accept evidence and interpretations that he cannot fully comprehend, he at least realizes that the ultimate 'authority' in science is natural phenomena, not the teacher or the textbook."[16] Statements such as this suggest that the leaders of CHEM Study would have been right at home in the late nineteenth century.

•   •   •

Of all the science curriculum projects of the postwar period, however, BSCS was the most radical by far in the way it used the laboratory with students. Given that the idea of scientific inquiry was most heavily emphasized in the biology project compared to those in physics and chemistry, it seemed a matter of course that Bentley Glass and his colleagues would push the edge of the laboratory experience envelope. Like the other projects, BSCS wanted the laboratory to do the heavy pedagogical lifting when it came to portraying the essence of science for students. "Students taking biology must be made to feel the elations, disappointments, the monotony and the hard work that are associated with research in science," noted one member of the Steering Committee. To this end, the group created a special subcommittee on laboratory innovation, and the main thrust of their work was developing what they called the "laboratory block" system, which entailed a number of intensive investigations that each would fully occupy the students over an extended period of time. It was an unorthodox approach, yet it ran into challenges that were not all that different from those encountered in the other curriculum projects.[17]

From the beginning, Glass and the lab innovation team sought to create authentic inquiry opportunities for students. They recognized that laboratory activities over the years had increasingly served to merely illustrate concepts, so much so that "students have come to spend most of their time watching demonstrations, looking through a microscope, dissecting animals or plants, learning names, labeling drawings—but rarely doing an experiment, in the sense of really investigating a problem, the answer to which is unknown." The lab team felt that the illustrative function of the laboratory could be better accomplished with good audiovisual aids matched to well-planned teacher demonstrations. This would free up space in the course for more in-depth inquiries that would "convey something to the learner of the nature of science, its methods and the spirit that pervades it." They worried too about school administrators who urged the elimination of laboratory work in response to "the predicted floods of students upon us, and the diminishing supply of qualified teachers." Proper laboratory work was nonnegotiable in their eyes. "Just as no amount of watching a skilled pianist playing the instrument can ever produce an understanding of what is involved so well as learning to play the piano oneself," Glass explained, "so in the scientific laboratory the novice learns for himself how to ask questions of nature and how to obtain unequivocal answers."[18]

In laying out what he saw as the two distinct functions of the laboratory, Glass highlighted an emerging shift in the nature of research in the life sciences. It wasn't that classrooms before BSCS had been misrepresenting scientific work in biology. They had, in fact, accurately captured the essential nature of that work prior to the early twentieth century. Biology (founded on the sciences of zoology and botany) had long had a strong descriptive, natural history focus. Understanding what the various forms of life were and their place in the natural world was the primary goal of naturalists. Glass recognized this, stating that the illustrative function of the laboratory "was probably the principal one in the minds of Thomas Henry Huxley and Louis Agassiz when they introduced it in biology." They had lived during a time, he noted, when "biology was for the most part still a descriptive science," and "their truth was a simple one: seeing is believing." But biologists in the postwar period, drawing increasingly on new techniques from the physical sciences, aimed to reposition biology as an experimental science on par with the higher-status fields of chemistry and especially physics. Thus the traditional descriptive exercises of the biology classrooms of the past were now seen merely as a pedagogical technique—something that could better be accomplished with audiovisual materials.[19]

The laboratory block program was designed expressly to engage students in the new experimental type of biological work. Contrasting it with what passed for hands-on learning at the time, Glass took pains to emphasize that it was definitely not "individual project work such as students prepare for science fairs."[20] The initial set of laboratory blocks included an exploration of microbes, a study of the relationship between structure and function in organisms, blocks on plant and animal growth and development, one on animal behavior, and blocks on genetic continuity, metabolism, and life in the soil, among numerous others. Each one required approximately six weeks of class time during which students would be deeply committed to considering the questions and exploring the material presented.

As Addison Lee, the chairman of the lab innovation committee, described it, "The student will study a given series of problems by making observations and obtaining data relevant to them. Then, he will be expected to interpret his observations and data and arrive at conclusions which will not only answer specific questions relating to the problem he has investigated, but also increase his understanding of the fundamental biological principles which underlie the individual Laboratory Blocks." The student was expected to make discoveries for him- or herself, and in doing so, the designers felt, "will generally follow in the footsteps of scientists who have preceded him," ultimately ending up at "the frontier of science." The lab group was certain that this process, in providing occasions for "the conduct of miniature but exemplary programs of enquiry," would ensure that students would "gain invaluable practice in working as a scientist."[21]

The block on microbes, for example, had students collect and cultivate samples of microorganisms from the soil and pond water using basic nutrient media. From the collection of organisms, which typically included various bacteria, protists, and fungi, students were instructed in how to produce a pure culture that could then be used for experiments. Over the course of the block, students tested their cultures to see what they required in terms of minerals, carbohydrates, and vitamins. They also tested to see how various pairs of microbes survived together in a given nutrient medium, and explored what nutritional deficiencies certain of the organisms might have. At the very end of the block, students were left with "unfinished" questions about why some samples grew and others didn't. Throughout this and all the other blocks, Glass explained during one of the Steering Committee meetings, they had tried to make "every question a question that could be answered on the basis of the student's observations. We emphasize the team work possibilities in the laboratory. Working in groups, one group would provide controls and show differences in results. We emphasize mathematical treatment of data. We insist that every measurement be made in the metric system and graphs be employed." And the early results were fairly positive. "We were told this was impossible for the high school level," he said, "but we felt the students were able to handle it."[22]

One of the essential elements of student work in the blocks was the whole-class discussion that came at the end. Teachers were advised in the laboratory manual to make sure this "summing up" happened. "Time should regularly be allowed for full discussion of each exercise and the questions it arouses," they were told. It could best be handled "in the form of a seminar, especially when one student has done an experiment different from the others, or a different part of an experiment." Once the student presented his or her findings, the authors wrote, the other students "can then discuss, question, and criticize." "This sort of training is invaluable," they insisted. During the "seminar" session, teachers were encouraged to model asking particular kinds of questions, such as "What were we trying to do?," "What data did we seek to gather?," and "What interpretations did we make of the data?" The goal was to "attempt to build an understanding of science as enquiry in a gradual way." One thing was certain: the teacher was essential to the success of this approach.[23]

In contrast to the PSSC lab exercises with their minimal direction and space for student experimentation and play, the BSCS laboratory block materials provided "highly structured" and "rigorously scheduled" activities over the course of a given six-week session. Careful planning was essential in creating the conditions for real inquiry to occur, according to Glass. For some biologists, though, the detailed instructions across the various laboratory exercises detracted from the authenticity of the experience. After looking over a laboratory activity on seed dispersal, one external reviewer wrote to the author, "This is a very good job" from a "teaching" perspective. But, he went on, "how about forgetting about those so-called research exercises which aren't really research and in fact are misleading students about what research is? . . . Stick to the game you know and be a champ." Despite such criticisms, the BSCS team had convinced themselves that they were on the right track. As Glass asserted, "Those visiting block classes can testify to the enthusiasm and the feeling students get for the nature of science investigation and inquiry. Careful planning and scheduling is necessary."[24]

•   •   •

Widespread classroom implementation of the BSCS laboratory plan was by no means easily accomplished, however. The biologists encountered a series of obstacles, some major and some minor, as the course moved from the planning stage to the test sites and finally out to the nation's classrooms. Many of those classrooms lacked the equipment necessary for the laboratory work envisioned. Years of emphasis on teacher demonstrations and student projects over the old-style laboratory work had by the 1950s left classrooms poorly equipped. "Often the existing biology laboratory is no larger than an ordinary classroom," observed two BSCS consultants in a report on the state of school laboratory facilities in the United States. "These rooms barely have more than one sink. . . . Basic necessities such as gas jets or electrical outlets are seldom provided for students. Ordinary classroom desks may serve as 'laboratory benches.'"[25] As troublesome as the equipment situation was, in the end it was the teachers—so accustomed to the traditional way of teaching biology—who posed the biggest challenge to the reform's success.

As the lab innovation group moved ahead with its work, the members knew that an influx of proper equipment would be essential for the inquiry activities to work as planned. Surveys of existing classrooms found that most either had little or no equipment or still retained classroom setups that were geared toward the "sterile kind of laboratory work . . . in which students are just labeling pictures." The passage of the National Defense Education Act provided the resources that were sorely needed. Money was soon flowing to districts ready to reinvest in science education, and BSCS was well aware of the potential. "The facilities and equipment," they recognized, "will reflect the kind of learning that is expected of the student." To help districts in the procurement process, the project developed a series of guidebooks and pamphlets for teachers, science coordinators, and administrators that described precisely how the NDEA Title III funding program worked and what equipment schools should consider as they moved toward widespread implementation.[26]

More than just the right equipment was needed, however. An ability to perform the various laboratory procedures with that equipment was required

Figure 19. Example of a "new" student lab from Biological Sciences Curriculum Study Bulletin No. 3 by Arnold B. Grobman et al., *BSCS Biology: Implementation in the Schools* (Boulder, CO: Biological Sciences Curriculum Study, 1964), 16. Report #BSCS-BULL-3. Copyright © by BSCS. All rights reserved. Used with permission.

as well. While it was certainly helpful to provide detailed directions for the students to follow in the blocks, the teachers themselves needed to master the new experimental techniques if they were to guide students in their inquiries. To address this, the BSCS group produced a series of "technique films" designed to show teachers how to perform the laboratory manipulations for each lab, such as using a microtome to make thin-section tissue preparations, separating mixtures by paper chromatography, and removing the pituitary gland of a frog. In addition, BSCS drafted supplementary teacher materials for each of the labs, published a detailed teacher handbook, and held summer training programs for those new to the lab block program as well as multi-week institutes for state educational agency supervisors and public school science supervisors. All these supports were designed to deal with what Glass and his colleagues recognized was "the somewhat complicated teaching approach" of the new materials, in order to "establish the general use of the Laboratory Blocks in American high schools."[27]

Despite the support and training provided, teachers still struggled. Among the issues that arose were the fact that too much material was expected to be covered in the time provided, the continuing challenge for teachers and students to perform the techniques asked for, excessively large class sizes (many teachers struggled with more than thirty students in lab settings), and, most significant, problems with guiding students through the reasoning process in their inquiries. Teachers were being asked to mentor students in a kind of work that they themselves had never experienced. "More instruction should be given to the teachers regarding the rationale of scientific procedure followed in the block," one report advised, "so that they may better interpret or explain instances of apparent failure or expectation of the experiments coming out as expected."[28] In other words, the teachers simply needed a better grasp of what reasoning through experimental research was all about.

At the heart of this issue was the fact that the research biologists were working with teachers who had grown up with the old-style biology; that is, they lived in a world where the descriptive, natural history approach of the past still held sway, laboratory activities were all about the observation and display of specimens, and identification and classification were the predominant intellectual activities. The new experimental biology of BSCS was foreign to them; it was a new way of thinking about what biology was, and they had to learn this along with the new techniques and experimental procedures. Glass and his colleagues knew that the transition wasn't going to happen overnight. "The teacher faces a rough road," acknowledged one member of the Steering Committee. "He is feeling the loss of security in the old way of teaching. Our courses are so new that he feels inadequate." They were hopeful that, over time and with continued professional development, the new approach would win out. Another member of the group predicted that "it will take four or five years for teachers to forget the myths about biology and the techniques they have used." In the moment, though, it seemed apparent that "the scientific inquiry aspect" of the course was "being fought by some of the teachers."[29]

Getting the teachers on board with the BSCS vision was difficult enough as it was. Complicating matters further were the actions of the various independent scientific supply companies that marketed and sold to teachers and school districts. The biologists were quite direct in the guidelines they provided schools for outfitting laboratories to handle the new inquiry approach. The lists included larger items such as refrigerators, autoclaves (for sterilization), incubators, kitchen ranges (for cooking bacterial media), and centrifuges, as well as balances, glassware, and assorted chemical reagents. This was the kind of material and equipment students would need to prepare the samples, conduct the experiments, and collect the data to be analyzed by their inquiry teams. Handy checklists were even printed up that a teacher could use to score how well outfitted a given classroom was to conduct the lab blocks. BSCS contacted the commercial suppliers to enlist their help as well in meeting the needs of the new approach. Ordering from suppliers and checking items off the purchase list should have been a relatively straightforward matter. But this wasn't always the case.[30]

Although many of the supply companies were glad to offer their services in support of the reform effort (especially given the potential for considerable profit from the flow of NDEA funds), others found themselves threatened by the shift to the experimental approach—particularly those that had long-standing relationships supporting teaching in the biological sciences. Ward's Natural Science Establishment, based out of Rochester, New York, was a classic example. Founded in the late 1800s by Henry Ward, the company began its life as a specimen supplier to the growing number of museums of the time. Ward spent most of his time collecting fossils, mineral samples, mounted animals, and skeletons to sell to institutions for public display. The company soon added products such as glass cabinets, display tables, specimen storage drawers, and other display-related supplies and

apparatus for museums and educational institutions to properly organize
and present their collections. Its entire business model, in other words,
was grounded in the natural history approach to knowledge that was characteristic of early botany and zoology. Everything Ward's sold facilitated the
"visual aid sort of objective" that BSCS was actively working to overturn.[31]

When the new experimental biology appeared on the scene with BSCS,
company directors at Ward's took notice. They had, of course, been informed
of the changes to come, as had executives at many of the other supply companies. For Ward's, though, with its strong natural history focus, this raised
concerns. "There is little doubt that these biologists are attempting to get
biology on an experimental basis in high school," wrote one executive.
"Whether or not they succeed is another thing, but at least they are going
to try." The problem was that product lines consisting of mounted specimens
and display cases were almost certainly going to be passed over as schools
adopted the new biology program. The shift to the biochemical and molecular aspects of the biological sciences embraced by researchers after the war
posed problems as well for the company. As Bill Gamble, the company director at the time, complained to colleagues, with respect to things such "as
DNA, genetics, molecular biology, and the like, at the present time we have
absolutely nothing to sell."[32]

The solution the company settled on was to develop new products that at
least appeared to be compatible with the new biology. They created a set of
what they called "curriculum aids," which were kits in which students were
given the materials and instructions for creating biological displays. One of
the curriculum aids, for example, had students take a biological specimen
such as a desiccated tarantula, pin it into an appropriate position, and then
embed it between layers of catalyzed bioplastic in order to create a hard trans-
parent plastic block through which all sides of the specimen could be observed. Other kits included materials for producing "temporary whole
mounts" on microscope slides and for injecting colored latex into the drained
blood vessels of various animals such as frogs, salamanders, and even small
alligators. When finished, these could be embedded in bioplastic as well, for
easy display. The challenge of developing products for the molecular biology
and biochemistry focus was met with the development of a series of transparencies that were designed to be displayed on the new overhead projectors
that were just coming into classrooms at the time. Visually depicting various
biochemical processes such as cellular respiration and DNA replication, the

transparencies—part of what Ward's called its "Dyna-Vue" system—came with printed handouts that students were to fill out at their desks as the teacher filled in the labels on the overhead.[33]

Classroom products such as these were well outside the bounds of what the BSCS project members would have accepted as appropriate. But in the intensive and often chaotic push for science education reform in the late 1950s and 1960s, it was often enough to market products as being aligned with the curriculum reform projects in some general way in order to secure their legitimacy. And this is just what Ward's did, selling the new kits and curriculum materials as an answer to the calls for more student hands-on work, for more student experimentation, and for engagement with the bio-molecular concepts that were being introduced into textbooks. Students using the new curriculum aids were without question actively engaged in novel hands-on activities, and such activities resulted in exciting products. What student, after all, wouldn't love to walk away from a school laboratory activity with a plastic-encased spider in his or her pocket to show friends and family? Such activities, however, did little to fulfill the learning goals BSCS had for students. They were activities that engaged students in technical work in a laboratory setting without involving them in the intellectual activity that characterized scientific inquiry.

Many of the new products from the scientific supply companies enjoyed tremendous success. Sales of the Dyna-Vue system alone generated more than $1.3 million in revenue over the ten years following its introduction. The success of these products stemmed from the shrewd calculation of company executives, who were able to market them as part of the new curricular reforms while at the same time aligning them with the traditional natural history approach to biology teaching—the production of specimens for display and the labeling of parts—that classroom teachers were most familiar with. In addition, the kit format provided teachers with activities that nearly ran themselves, which was a welcome relief for those who struggled to manage teams of students engaged in sophisticated, open-ended inquiries that often led to ambiguous results that they struggled to make sense of. The commercial success that a company such as Ward's enjoyed during the NDEA funding era, in other words, came at a cost to the student understanding of scientific process envisioned by the BSCS team.[34]

•   •   •

In spite of the short-term challenges they faced, the new curriculum projects rode a wave of public and political support into the nation's science classrooms. There seemed to be no alternative, given the national security threat posed by the Soviet Union and the high status districts enjoyed as they adopted materials authored by world-renowned scientists. Who, after all, knew more about science than this collection of research stars and Nobel laureates? By the early 1970s, well over 60 percent of all high school students were taking courses that were using one or more of the new NSF-funded curricula. This was true in over 40 percent of biology classrooms (with BSCS materials), 30 percent of chemistry classrooms (with the CHEM Study program), and 30 to 35 percent of physics classrooms (with the PSSC course). In an attempt to stay relevant, existing commercial textbooks shifted their emphases to follow the lead of the NSF curriculum materials, resulting in far more students getting a taste of the new approach. Between 1964 and 1974, print references to the phrase "inquiry approach" spiked, increasing sixfold over that ten-year period. Despite the apparent success, the real question of how faithfully teachers were able to implement these reforms with their students remained unanswered.[35]

In the mid-1970s, after twenty years of active curriculum development and implementation, the National Science Foundation conducted a thorough review of the status of what by then had come to be called the Course Content Improvement program. The coordination and oversight of the review was handled by F. James Rutherford, the assistant director for science education at NSF.

Rutherford was a former California high school physics teacher who had gone on to earn his doctorate in education at Harvard in 1963 under Fletcher Watson, a science education professor who had worked closely with James Conant on his general education initiative in the late 1940s and early 1950s. Rutherford actually began his graduate studies in the history of science under I. B. Cohen at Harvard in 1954 (though he never completed a degree) and studied as well with the physicists Edwin Kemble and Gerald Holton (who developed the Nat. Sci. 2 course for Harvard undergraduates that focused on the development of science concepts in their historical context) before turning formally to science education. Drawing from this experience, Rutherford joined with Holton and Watson in the 1960s to develop a high school version of the Nat. Sci. 2 course as a more humanistic alternative to PSSC. That course, Harvard Project Physics, was made available for widespread classroom use in 1970. Having worked on NSF-funded curriculum reform himself, albeit in the second rather than first wave of projects following Sputnik, Rutherford was well aware of the implementation challenges reformers faced.[36]

The NSF review consisted of three main components: an extensive literature review, a national survey of public and private school teachers, and a series of in-depth case studies of various programs and institutions located everywhere from the Pacific Northwest down to the South and across to the eastern seaboard. Collectively the studies, which were conducted over the years 1976 and 1977, provided an exhaustive look at the fate of the science curriculum reform effort of the time, including not only whether the courses were formally offered in high school classrooms but also, and more important, just how the curricula were used. The detailed case studies of eleven schools—rural, urban, and suburban, large and small—were particularly revealing. Rutherford wrote in the foreword to the report that "the two decades of activity in science and mathematics education that began in the mid-fifties are unique in the annals of American education. Under the leadership of the National Science Foundation scientists and educators were mobilized in an unprecedented effort to strengthen science and mathematics learning." But then he posed the critical question: "What was the lasting outcome?"[37]

The NSF evaluators had little trouble finding evidence of the rhetoric of reform in research publications of the time and among the survey and case study data. There was, for example, "widespread agreement in the literature about the importance of science laboratories and facilities to the science program," and across the variety of legislative and policy documents of the period there was a clear emphasis on inquiry teaching and the investigative process as a "basic and continuing need" in science education. This had been the focus of science teaching over the past twenty years, asserted the Stanford science education professor Paul Hurd in 1976. But a closer look on the ground revealed that although the various report findings couldn't be generalized to all cases, they were disappointing and troubling all the same.[38]

Teachers enacting true inquiry-based instruction as envisioned by the scientist reformers were, it turned out, difficult to find in everyday classrooms. "In the opinion of many enthusiasts," evaluators found, "these programs were too demanding of teachers, requiring a rather drastic shift in ways of thinking about teaching as well as changes in teaching practice." Some teachers felt that students were capable of learning via inquiry, but few "were confident they could make *systematic* inquirers out of them. It took[,] they said, a special kind of faith, a special patience, to draw out and refine what the traditions of schooling did not encourage."[39]

What the evaluators did find was that "despite considerable contact with legacies of the NSF-sponsored curriculum projects and with inservice programs dedicated to the promotion of student inquiry, very little inquiry teaching was occurring in science, math, and social science in the eleven sites." Some classroom teachers had never even heard of the new materials and approach. One site visitor reported being "amazed at the number of teachers who were not aware of the curricular developments of NSF and other groups." And countless teachers, whether they were aware of the projects' origins or not, were simply not using the materials, despite the hundreds of millions of dollars spent developing and testing them and then shipping them to schools. "Many of the materials developed to promote inquiry in children . . . were still there in the schools, often in storage or stacked in corridors, seldom being used."[40]

Even the more traditional laboratory work, in which students followed a prescribed sequence of activities, seemed scarce. Some reports found that hands-on activities were more common than they had been prior to the appearance of the NSF-funded projects, possibly due to the appearance of the kits and instructional materials put out by companies such as Ward's. But when evaluators asked high school seniors from twenty-eight districts around the country for the one thing that was "most wrong" about their current science courses, 43 percent indicated that the courses either "overemphasized facts and memorization" or did not have "enough lab and project work." That view wasn't generally shared by the teachers, however, nearly two-thirds of whom felt that the balance between textbook work and laboratory and project work was just about right. Those views notwithstanding, "laboratory work in several sites appeared to be diminishing in importance," according to the evaluators, and they attributed this to "the expense, vandalism and other control problems, and the emphasis on course outcomes that would show up on tests." Another factor may have been difficulties in securing necessary equipment. There were huge variations in the amount and kind of apparatus and materials available. Depending on the district, evaluators described the situation as feast or famine. "In half the high schools laboratory science was reported to be nearly impossible to conduct because the labs were run down or ill-equipped," despite the resources provided by the NDEA to address this very issue.[41]

If students weren't engaged in learning science through inquiry or laboratory activities, then what did science instruction look like in the mid-1970s? It appeared to be little different from the rote learning that had been common in the nineteenth century. An episode of teaching captured by one of the site visitors to a classroom in Pine City (a pseudonym) was deemed typical. The scene involved a teacher standing before a large group of students asking them questions:

"What are the three characteristics of the nervous system?"

"What's the difference between a threshold and subthreshold stimulus?"

"What's the difference between the nervous system of the amoeba and the human?"

According to the visitor, "the answers come back in the stylish rhetoric of the *textbook*. Clearly the essence of the task has been to search the text for the sentence which contains the correct answer." Summing up the situation, evaluators wrote, "Teachers relied on, teachers believed in, the textbook. Textbooks were not used to support teaching and learning, they were *the instrument* of teaching and learning." They added that "*information* is pretty much what many of the courses are about."[42]

This applied to the "five steps of the scientific method" as well, which (along with other elements of "old science" such as "the three classes of levers" and "the stages of mitosis") had "been in the school curriculum a long, long time—fifty years at least." Ideas such as these continued to be taught more as a nod to tradition than anything else. "Their existence in the curriculum," the evaluators commented, "was easily rationalized in terms of knowing what people are talking about when they refer to these things they studied in school." In other words, for these teachers the scientific method was something students just had to learn, by rote if necessary. As for the new post-Sputnik approach to science education, one of the NSF review authors characterized the situation rather bluntly: "Inquiry does not appear to work."[43] After eighty-some years of various reforms and the investment of hundreds of millions of dollars, the schools didn't appear to have gotten any closer to teaching students to understand in any meaningful way the methods by which science operated.

9 Project 2061 and the Nature of Science

IF THE PERIOD from the mid-1950s through the 1960s was the golden age of science education, with its unprecedented levels of federal funding and involvement of Nobel laureates and other top scientists, the decade of the 1970s into the first half of the 1980s represented an era of comparative neglect. Attention to academic subjects in schools declined with the new political emphasis on urban poverty and concerns about the repressive nature of formal education that arose in the late 1960s with the more liberal views of the role of schooling in American society. At the same time, the image of science fell from public favor in these years as a result of its association with environmental degradation, on the one hand, and with militarization and the Vietnam War, on the other. This shift in educational priorities along with the new critical views of science led to conditions of general decline in science education that raised alarms among the scientific elite.[1]

Those worries were felt acutely by the leadership of the American Association for the Advancement of Science, prompting its executive director, William Carey, to recruit Jim Rutherford in 1981 to lead a new effort to rebuild science education across the nation. Rutherford came to the AAAS from his stint as assistant director for education at the National Science Foundation via a position as assistant secretary for research and improvement at the newly established Department of Education. He was someone familiar with the levers of change, such as they were, in the American educational system. Rutherford's immediate task was to move education to the top of the AAAS agenda, and specifically to enact elements of a January 1981 resolution passed by the association's board of directors "to reverse the damaging decline of science and engineering education in the United States."[2]

The challenge Rutherford faced was formidable. Since the mid-1970s, science education had been pushed to the margins of public consciousness, and a crushing recession along with the education aversion of the newly installed Reagan administration made prospects for any federal initiatives bleak. "It is easy enough to say that business and industry, the scientific and engineering societies, and the foundations ought to pick up the slack," Rutherford wrote to a friend. But it wasn't clear to him at the time what those institutions could do to really make a difference. What was obvious to Rutherford was that a long-term plan was needed rather than some quick fix. As he saw it, his job was to develop something that "the association can stick with for the decade or longer that it takes for anything to have a lasting impact on our complex educational system."[3]

Rutherford mulled the possibilities in the summer of 1982. The scientists of the 1950s had had the shock of Sputnik and the military threat from the Soviet Union to help usher reforms into the schools. The biggest threat of the early 1980s, however, was economic—from Japanese automobile imports, for example. Grasping for something bold and symbolic, Rutherford latched on to Halley's Comet—a satellite of a different sort. It seemed to fit the bill. The famous comet, he noted, was due to appear that October, the same month as the twenty-fifth anniversary of the Sputnik launch, and it would be seventy-five years before it would return again. What sort of changes might take place in our civilization between those visits? mused Rutherford. Looking at the dramatic changes that had occurred between prior fly-bys, he concluded that "we cannot accurately describe the world as it will be when Halley's Comet next returns." However, he asserted, we do know that "the changes that will be brought about in our culture, in our way of life, will have more to do with the utilization of science and technology than with anything else."[4]

What was needed, in Rutherford's view, was an entirely new approach to science education, one that would prepare children born in 1986—the year the comet would make its closest pass by Earth—to live in the scientific and technological world to come when those students would grow up to work, have children, eventually retire, and live to see the comet return in the year 2061. The length of time between sightings gave Rutherford the long view he and the AAAS were looking for to avoid yet another crisis-driven crash program that was unlikely to produce meaningful and enduring change. Project 2061, as he named it, was bold and imaginative, clearly something outside the typical educational-reform box. Its central goal was to articulate "what understanding of science and technology will be important for everyone in tomorrow's world" and then to work toward realizing that goal in a systematic way.[5]

The vision was formally presented in 1989 with the publication of the Carnegie Corporation–funded report *Science for All Americans*. It centered on an idea of science literacy (a concept first introduced in the late 1950s) that fused science content knowledge with science methods. In other words, "knowledge" was conceived as an overarching concept that included both "knowledge of the conclusions of science and of the processes of generating scientific conclusions."[6] More significant was the project's effort to define not just the essential nature of science but also how it functioned institutionally and how it might be applied to societal concerns in an era where the boundary around what counted as science was not always clear. Public understanding of science methods was crucial to how that boundary would be determined.

●   ●   ●

The status and prestige of science were hit hard by the cultural ferment of the late 1960s and early 1970s. The frustrations with establishment ideologies that erupted on college campuses gave rise to a countercultural movement that railed against technologies of engineering and social control. Humanists joined with protestors of all sorts to challenge the dominant role science was playing in the corporate-military machine. The movement soon spread beyond the campus and was taken up by segments of the mainstream middle class as well. Many of these individuals looked for alternatives to what they saw as the dehumanizing and dangerous influence of science, embracing mysticism, mood-altering drugs, environmentalism, and other nontraditional worldviews and practices. Survey data in 1971 showed that only about a third of the public had "great confidence" in scientists, which marked a new low. And although public attitudes rebounded somewhat by the mid-1970s, pseudoscientific ideas about things such as astrology, biorhythms, and paranormal activity seemed to be on the rise nonetheless.[7]

In this era of skepticism, many in the scientific community worried about the willingness of the public to adequately support a robust research enterprise in the United States. Moreover, reports of declining student test scores cast doubts on the quality of those entering the science-workforce pipeline needed to maintain the country's research excellence. This was all the more concerning to business leaders and policymakers, who increasingly believed that science and technology were central to economic innovation. In an era of new global competition and declining economic performance, science was vital, and the neglect of coordinated efforts to cultivate public understanding of science, many felt, was a recipe for disaster.[8]

It is nearly impossible to think about the 1970s without economic anxiety coming to mind; in fact, the phrase almost defines our understanding of that decade. The OPEC oil embargo of 1973 and the energy-price shocks it triggered pushed the United States into a severe recession that lasted nearly two years. The stock market plunged and then, after a brief recovery, followed a long downward trend through the early 1980s. American productivity dropped by half compared to the decades immediately after World War II, and unemployment rose to an average of 7.4 percent for most of the decade, partly a result of jobs moving overseas. High unemployment teamed with high inflation—averaging over 8 percent annually from 1973 to 1983—to create what came to be called a period of economic "stagflation." A second oil crisis in 1979 (which doubled the price of crude) produced long lines at gas stations and further rattled the American public. Adding to the troubles, the ongoing instability in oil supplies created a market for fuel-efficient foreign automobiles from companies such as Datsun and Toyota, which depressed domestic automobile sales and led to additional job losses. Newspapers talked of the growing Japanese "invasion" or "challenge." Under such circumstances, few could find much reason for optimism.[9]

During the depths of the recession, innovation in science and technology was floated as a way out of the country's economic misery. The competition from Japan's consumer electronics and automobile exports seemed to point the way forward. Academic researchers and government policy analysts began studying the relationship between technological breakthroughs and the economy early that decade, and soon industry leaders started calling for greater attention to what they saw as a growing innovation deficit. By the late 1970s, magazines such as *Time* and *Business Week* were highlighting the need for government policies that would promote innovation, and an article in the *Washington Post* asked outright whether the United States was "losing its edge in technology" compared to the rest of the world. As the decade drew to a close, an increasing number of economists, policy experts, and government

Figure 20. Cover of *Newsweek* from June 4, 1979. Reproduced from a copy in the author's collection.

leaders agreed that industrial advancement driven by new technologies would be the key to sorely needed economic growth.[10]

While the business community began to awaken to the potential of science and technology in the early seventies, the scientific research establishment, to its dismay, experienced continued hits to its status and influence among the broader public. As the decade opened, Congress cut back funding for scientific research as questions surfaced about the practical relevance of ivory-tower inquiries. And in response to unwelcome advice on national security matters and a perceived left-leaning political bias, President Nixon banished the scientists. He declined to award any National Science Medals in 1972, and in 1973 he abolished the President's Science Advisory Committee, the Office of Science and Technology, and the position of special assistant to the president for science and technology. These moves severely damaged the close working relationship that had been forged between scientists and the executive branch during the Eisenhower and Kennedy administrations and left the "scientific-technological élite," in the words of one observer, "radiating the mood of a *déclassé* set awaiting the next disaster."[11]

More broadly, the growing public distrust of university and government experts in the United States seriously undermined any centralized social programs and initiatives. This shift in public sentiment effectively curtailed the NSF-funded education programs that had grown up since the 1950s. Religious groups organized in the early seventies to push back against the reintroduction of evolution into schools that came with the publication of the BSCS textbooks. At the same time, what came to be known as the "new math" curriculum, which famously presented the subject in a way that few parents could easily grasp, suffered under attacks that quickly led to its demise, replaced largely by a grassroots "back-to-basics" push in education that stressed doing math to get correct answers rather than to achieve the deep conceptual understanding and appreciation of the discipline that reformers were after.[12]

The death blow to the NSF curriculum reform movement came with the public controversy over the middle school social studies project "Man: A Course of Study" (MACOS), a curriculum that sought to have students learn about both human and nonhuman behavior through the eyes of the social scientist. Parents and others objected that the materials, among other things, promoted cultural relativism, denied humans' special place in nature, and engaged in social engineering that undermined traditional American values.

A series of congressional hearings and audits of this program and then of all NSF education initiatives resulted in the termination of nearly all funding in 1976. (It was this intense scrutiny that led to the evaluation studies conducted with Rutherford's supervision in 1977.) The people had spoken, taking the academic elite down a peg or two in the process.[13]

After the outright hostility of the Nixon administration, scientists were hopeful that the election of Jimmy Carter in 1976 would usher in a more favorable climate for science. Trained in nuclear engineering while in the navy, Carter recognized the importance of science to national strength. He highlighted the role of research and development in his 1978 State of the Union message to Congress and dispatched Frank Press, the director of the newly formed Office of Science and Technology Policy, to address the AAAS directly at its annual meeting that same year. In his remarks, Press aligned with the emerging consensus that the country was "locked into a dynamic system of global economic growth, . . . one based largely on technological change and innovation," and promised that a funding renaissance was on the horizon. Indeed, federal money for basic research and development increased more than 30 percent during Carter's time in office. However, the landslide election of Ronald Reagan in 1980, who campaigned on radically scaling back government programs and slashing what he saw as a bloated federal budget, showed how fleeting political favor could be. Rough times for science again lay ahead.[14]

●   ●   ●

The change in administrations in 1981 profoundly altered Rutherford's position in the science education world. Whereas previously he had worked as a federal government insider, a political appointee striving first to rebuild the education efforts of a politically hobbled National Science Foundation as assistant director and then to manage education research in the Department of Education, now he was on the outside, watching a man take office who vowed to eliminate the department, wipe out science education at the NSF, and shift nearly all responsibility for education to the states. While on vacation in Mexico that winter, he considered his options and, while doing so, reread an old copy of the Conant-initiated Harvard report, *General Education in a Free Society*. This report had advanced a vision for education devoted to fostering broad public understanding of what science was and how it worked, providing one of the first statements of "science literacy" before the phrase had been introduced and popularized. This reading, as he recalled years later, rekindled his "determination to continue in science education." The day after Reagan's inauguration, he took his position at the AAAS and began laying out plans for what would become Project 2061.[15]

It was clear to him at the outset that forces outside government would need to be mustered to make any real progress. Reflecting on the stance of the new administration, he explained that "the federal government is now in retreat educationally, if not in disarray. The President of the United States has made it clear that whatever the problem is, it is a matter to be addressed by the individual states and local school districts." This retreat, he insisted, made it "more urgent than ever that AAAS provide national leadership in science education." Rutherford began courting the private foundations to provide the resources needed for the work that lay ahead. He had firsthand experience with the power of philanthropic institutions to make things happen. Harvard Project Physics never would have gotten off the ground without the support it received from the Carnegie Corporation of New York. So in his new role at AAAS, Rutherford almost immediately started conversations with key figures at that foundation to see what might be done to bolster science education at this crucial juncture.[16]

Officials at the Carnegie Corporation were receptive to his overtures. The organization routinely supported projects in science and education. Before its support for Project Physics in the 1960s, Carnegie had provided nearly all the funding for Conant's 1949 Conference on the Place of Science in General Education as well as for the subsequent summer sessions with teachers convened at Harvard in the early 1950s. The Carnegie Corporation was aware of the growing concern about education and its role in economic development. Numerous private and public committees and commissions were developing various reports on the condition and future direction of education in the United States in the early 1980s, and Carnegie hoped to leverage the public attention these reports would generate, to take "advantage of the current momentum to develop a coherent national education policy linked to economic development." "The economy," Carnegie officials felt, "could be the modern Sputnik, a powerful lever for reform and support." At base, officials believed that "the United States and other advanced industrial nations [were] undergoing a long-term transformation driven by scientific and technological change" that would require "the nation to upgrade substantially the education and training of the workforce."[17]

Among the many reports in 1983, none influenced the public debate as much as the publication of *A Nation at Risk* by the National Commission on Excellence in Education, chaired by David P. Gardner, president-elect of the University of California. The commission was formed at the request of Terrel Bell, the new secretary of education under Reagan, to take stock of the state of public education at the time. The report, like many of the others, noted the coming knowledge economy and warned that the United States was ill-prepared to meet the challenge. Pointing out the troubling international test performance of American high school students and the continuing decline in academic achievement and national standardized test scores, the commission called for increased rigor across all the core academic subjects in high school.

The publicity surrounding the publication was due as much to its alarmist tone as it was to the subject on which it reported. The opening paragraphs spoke of American society being "eroded by a rising tide of mediocrity that threatens our very future as a Nation and a people." It went on to claim that "if an unfriendly foreign power had attempted to impose on America the mediocre educational performance that exists today, we might well have viewed it as an act of war." Jarring words, to be sure, but they grabbed the public's attention and ultimately had a very real effect. David Hamburg, president of the Carnegie Corporation, found it "hard to believe that education could move so quickly to the top of the national agenda."[18]

Across nearly all the reports were two recurring themes. The first was, naturally, the driving focus on the connection between education and economic growth. In an earlier, Carnegie-sponsored meeting on the topic of education and the economy, the question was posed whether a substantially increased "investment in education for all" would result in those costs being "offset by increased output in the economy." The other theme was the emphasis on reforming the educational experience for "all our future citizens" rather than "just for the self-selected elite" (as was perceived to be the case with the Sputnik-era reforms). The concern to create a new science education for the general population of students was reinforced with the publication of the spring 1983 special issue of *Daedalus* dedicated to the topic of "scientific literacy"—a phrase that quickly became the tag line for science-education-for-all goals in the late twentieth century.[19]

While economic arguments predominated among the policymakers and business leaders, the AAAS and Rutherford aligned themselves with the "science-for-all" camp. "All" was "specifically meant to include the children who have heretofore been shunted aside because of race and gender." The inclusive nature of the project aimed, in particular, at the role of science in citizenship, which they believed to be the bedrock on which any reforms should rest. Rutherford readily admitted that the problems of economic productivity and even military preparedness required "achieving a high level of scientific/technological literacy among decision-makers and workers," but he insisted that having "an elite, highly trained cadre of scientists and engineers is not sufficient." Overlooked was the fact that "it takes a large number of well educated people to operate a technology-based democratic society," including women and minorities, who had "been largely by-passed by science and engineering." Without a major initiative in science education, "the gap between the public understanding of science and technology and the requirements of citizenship in a participatory democracy will continue to widen."[20]

In his communications with the Carnegie Corporation, he pushed hard on this point. When Alden Dunham from Carnegie asked him for feedback on a draft report of the new program it was launching, "Science, Technology, and the Economy," Rutherford wrote back asking immediately about the title: "Why the 'economy'? It makes it sound as though the purpose of your programs are only to revive the economy. . . . I do think you have broader goals in mind than the economy. Instead of 'the economy,' would it be more valid to use 'the citizen'?" The "citizenship argument," he explained, is "a more powerful one than the economic argument."[21]

Although Rutherford failed to convince Dunham to change the name, the two appeared to agree that whatever economic needs might be served by the new education program (they admitted that the connection between education and the economy was never all that clear), raising the level of scientific literacy in the United States would be a central focus. As the AAAS asserted in its own "crisis" report at the time, which Rutherford had a hand in drafting, "The decline in the public's science literacy may be more injurious to the nation in the long run. Our future citizenry as a whole will be even *less* well prepared to understand and support scientific development." This was a "citizenship" argument of a particular sort. In its reference to generating "support" for scientific work, the AAAS was self-serving, to be sure. But in this situation, everyone seemed to get what they came for. In June 1985, officials at Carnegie approved an initial grant of $923,196 for the first phase of Project 2061. It was to be Carnegie's "flagship project."[22]

The goal of Phase I was to "crystalize out of three centuries of modern science and many millennia of technology" the essential knowledge that all citizens, *"regardless of their vocational aspirations,"* should know by the time they finish high school—to determine "what science is most worth knowing" (as early drafts of the proposal were titled).[23] Here Rutherford invoked Herbert Spencer's 1860 essay "What Knowledge Is of Most Worth?," which itself served as a prominent justification for bringing science into the formal school curriculum in the nineteenth century.

Rutherford and his staff organized the work of the mid-twentieth-century reformers across five categories: the physical and engineering sciences, the biological and health sciences, the social and behavioral sciences, technology, and applied mathematics (as they were initially labeled). Panels of distinguished scholars—mostly academic scientists—were invited to serve in each area. "Teachers and other educators" were left out of the first phase, given that Rutherford felt they were "not in a position to know what constitutes 'good' science." From the five panel reports, an oversight committee, the National Council on Science and Technology Education, was to draft a single, consensus report—something that was eminently readable, with some style and panache—detailing "what knowledge, skills, and habits of mind associated with science, mathematics, and technology" all Americans should have by the time they leave school.[24]

The project altogether was envisioned to span three phases. The first focused on the formulation of the essential knowledge of science; this was followed by Phase II, which entailed developing "educational guidelines for sustained and purposeful reform" in consultation with educators and education researchers; and the final phase centered on the oversight and implementation of systemic reform. This was no small undertaking, and Rutherford gave the project plenty of lead time—until 2061—to achieve its goal. Even with that time span, one reviewer commented, "it still boggles the mind." AAAS director William Carey wrote to Rutherford that he read the proposal "with bulging eyes and a few tremors." "It's a prodigious undertaking," he commented, "probably the most ambitious that AAAS will have attempted," but one he also felt it was imperative to pursue. By the summer of 1985, that undertaking had been launched.[25]

•   •   •

As the panels got down to work, they didn't focus on defining the essential process or method of science in any deliberate way. This neglect was likely due to the manner in which their task was framed—which was to determine the essential "knowledge" of the scientific domains in question. Instructions to the biology panel, for example, began, "The goal of your panel is clear enough: it is to identify what knowledge within the entire sweep of the biological sciences is most worth learning by all young people." Process wasn't part of the charge, at least not explicitly, which worried some of the early reviewers. One felt that this focus would result in a view of science as "a vast body of knowledge somehow to be curricularized and syllabified" rather than as a "cultural enterprise."[26]

Rutherford had always thought of scientific knowledge as including both "knowledge of the conclusions of science and of the processes of generating scientific conclusions." He hewed close to the views on this articulated by Schwab. One of the main problems in science education, he wrote early in his career, was "a failure to recognize and take fully into account the close organic connection between process and content." While he shared this nuance at one of the early meetings, it was often overlooked as the panels put together their reports. Talk of the scientific method came to the table nonetheless in a number of ways, and a description of how science worked made its way into the final report in a separate chapter Rutherford wrote himself. This focused more on describing what science was, though, than on its method.[27]

Throughout their discussions, panel members routinely used "scientific method" as a shorthand to talk about the process of science, despite the efforts of Conant and other postwar scientists in the 1940s to beat back the idea that such a thing ever existed. At one of the physical science meetings, George Bugliarello, the panel chair and president of the Polytechnic Institute of New York in Brooklyn, insisted that "people coming out of high school must know something about . . . the scientific method" and why it is important. Another panel member similarly tossed off the phrase: "Somewhere, in talking about the scientific method, we have got to say not only what it can do, but what it cannot do." At times it was defined explicitly as a form of logical reasoning ("induction, deduction, other aspects of scientific logic"), as the process of "observing phenomena and drawing inferences; designing experiments and predicting outcomes; revising theory from evidence," and so on. But at other times it was cast as a method of thinking more generally, not much different from how Dewey had explained it in the early 1900s. Among the panel members, according to meeting notes, "there was agreement that the process of 'how to think' should be taught" and applied to "everyday human experience."[28]

The identification of scientific process with a generalized method of thinking fit well with the project's emphasis on science literacy. Such thinking, the group believed, would be useful to students in confronting the challenging science-related issues they would face as adults. Records from the meeting of the biology panel, for instance, showed that the group believed a focus on "scientific methods of inquiry" was "fundamental to being an informed citizen regarding future 'crises,' 'technological fixes' and scientific breakthroughs." This sentiment was echoed time and again as the group hammered out their reports. During a meeting of the oversight council in 1986, one member said this explicitly. "It seems to me," she stated, "that what you want to do is teach kids how to think." "In science education," she went on, it is "a particular way of approaching problems . . . so that they can participate as citizens in a democratic country and make informed choices." The biologist John Moore, who had worked on the BSCS curriculum materials earlier in his career, noted that "human beings make human decisions, for whatever reasons. But at least the sciences can give them a better basis for reaching what is hopefully a rational decision." Panel members certainly felt that much was at stake. Gerard Piel, the longtime publisher of *Scientific American,* emphasized the seriousness of their work: "That spirit of what the obligation of citizenship is, on the one hand, and the liberating, . . . empowering of the individual that education in science will give," on the other—"these are the grounds on which we may urge the importance of our mission here."[29]

The imprecise definitions of and vague sentiments about method notwithstanding, *Science for All Americans* did explicitly address what science was and how it worked. Rutherford organized the opening chapter, titled "The Nature of Science," into three sections, covering the scientific worldview, scientific inquiry, and the scientific enterprise. Across the first two sections he sought to convey the essential elements that distinguished science from other social and intellectual activities. He explained that science assumes a world that is understandable and that scientific explanations and ideas, while durable, are also subject to change. In the section on inquiry, Rutherford offered no pat explanation for how science as a singular endeavor generated new knowledge. Taking up Schwab's appropriation of Dewey, he wrote that "scientific inquiry is not easily described apart from the context of particular investigations." And, following Conant and the postwar anti-method crowd, he emphasized that there certainly was "no fixed set of steps that scientists always follow, no one path that leads them unerringly to scientific knowledge."[30]

He did, however, identify key features of inquiry that were "especially characteristic of the work of professional scientists." These included the idea that "science demands evidence," that "science is a blend of logic and imagination," that it "explains and predicts," that "scientists try to identify and avoid bias," and that "science is not authoritarian." Under each of these headings, Rutherford explained what he meant by the feature in question and gave examples to illustrate it. For "science is a blend of logic and imagination," for instance, Rutherford wrote that "scientific concepts do not emerge automatically from data or from any amount of analysis alone. Inventing hypotheses or theories to imagine how the world works and then figuring out how they can be put to the test of reality is as creative as writing poetry, composing music, or designing skyscrapers."[31] Perhaps in line with the larger report's focus on "knowledge outcomes," these features were more about what science was than how it operated in practice. This was a departure from the approach taken by the reformers of the 1950s and 1960s who were concerned primarily with students engaging in the process of inquiry firsthand or understanding knowledge about science implicitly from laboratory activities or textbook narratives.

The last section of the chapter covered what Rutherford called the "scientific enterprise"—a topic he touted as something "entirely new" in the realm of school science education. As he explained in one internal document, it wasn't enough for students to know the key concepts of science, "nor is knowledge of the so-called 'scientific method' sufficient, for that deals with the idealized logic of investigation, not with the internal operations and social structure of the entire system of science." "For the purposes of citizenship," he insisted, "some grasp of science as an enterprise is essential." In this

Figure 21. F. James Rutherford (right) at Project 2061 roundtable discussion at AAAS Annual Meeting, Chicago, February 1987. AAAS Project 2061 Files, Box 7. AAAS Archives and Record Center, Washington, DC.

section he highlighted, among other things, the social nature of scientific work, how science is organized into disciplinary fields, and who scientists are (calling particular attention to women and minorities) and what sorts of institutions employ them. Finally, and perhaps most salient, he ended with a discussion of how research is funded and by whom, as well as who decides which questions are pursued and which are not. This focus on the social, political, and institutional infrastructure of science was, without question, a novel feature of the AAAS report.[32]

The multifaceted presentation of the nature of science in *Science for All Americans,* in terms both of its more sophisticated treatment of knowledge production and of its inclusion of the political economy of the research enterprise, drew from new understandings of science that had emerged from the scholars who began to study science and its operations in the years after World War II. The government-fueled expansion of science and its increased entanglement in security and societal affairs in that era prompted researchers to explore the conditions of science and its growth. The Harvard general education group was a central player in such work. The efforts of an emerging generation of scholars housed at places such as Wisconsin, Cornell, and the University of Pennsylvania in addition to Harvard were encouraged by the National Science Foundation through the establishment of its History and Philosophy of Science Program in 1957. More "science studies" departments, employing philosophers and sociologists of science in addition to historians, were soon established at universities in the United States and the United Kingdom. Scholars in all these areas built an entirely new body of literature on the social and institutional organization of science, the historical origins of scientific thought and practice, and the nature of scientific progress.[33]

Rutherford's views of science undoubtedly were informed by the intellectual atmosphere of Harvard, as were those, though less directly, of the many science educators who came of age with the flourishing of science studies. As a science teacher in California in the 1950s, Rutherford was drawn to the argument for science literacy put forward by the Redbook report and had familiarized himself with Gerald Holton's ideas about science education as well as Conant's efforts to remake the undergraduate curriculum at Harvard. One might say Rutherford had been completely immersed in the history and philosophy of science as it came of age in Cambridge in those early years, and it wasn't long before he teamed up with Holton and Watson to work on Harvard Project Physics. The intellectual bond Rutherford established with Holton (who played a key role in drafting the *Nation at Risk* report), in particular, was lasting and something he clearly valued. When it came time in the 1980s to put together the plan for 2061, Holton was one of the first people Rutherford tapped to advise him on the project.[34]

The portrayal of the nature of science in *Science for All Americans* reflected the newer thinking coming from science studies at the time. Historians and philosophers of science were, in fact, involved in the report's development, and panel members drew on Thomas Kuhn's writing about paradigms and scientific revolutions at various points as they formed their ideas about how science worked. Despite these influences, however, the final image of science presented was far from revolutionary, and this was primarily due to the purpose the project served, which was to bolster the science establishment during a time of sociopolitical vulnerability. Through the "entirely new" material on the scientific enterprise, for example, 2061 sought to educate the public about the importance of institutional support, peer review, and scientific autonomy. Responding to a statement Rutherford made in an early draft that the support of science should not "be left up to the scientists themselves," one reviewer wrote that "many readers will assume you mean not only the overall level of support, but also the division of funding within science." He went on to remind Rutherford that "heroic battles have been fought to assure that it is peer review that determines this, and in the instances where that process has been overtaken by nonscientific viewpoints . . . the result has been a disastrous waste of funds. Please delete!"[35]

●   ●   ●

Project 2061's emphasis on the "nature of science" as a broad category that spanned ideas about science and the process of inquiry wasn't a completely new way to think about what the public should know about science. Researchers in science education had used the phrase in the 1960s to think about student learning as the field sought to develop ways to measure the more expansive kinds of understanding that might come from historically based courses such as Conant's. But Rutherford placed the nature-of-science construct front and center in Project 2061's pursuit of science literacy, and its use was particularly effective for AAAS's purposes at the time. Not only did it allow the project to avoid the content / process dualism that he felt misrepresented the holistic nature of scientific work but also, and more important, it provided a way to use the project to protect certain areas within science—the social sciences and evolutionary biology, for example—from external threats to their legitimacy and broader cultural authority in the 1980s. The appeal to a common, yet distinctive methodological approach was key in both instances.[36]

The social sciences—disciplines such as anthropology, sociology, economics, and psychology—had always had an ambiguous status as science in the United States. Prior to the founding of the National Science Foundation in 1950, there was considerable debate within Congress over whether they should be included for funding alongside the natural sciences, and they were not. The door was left open for NSF to bring them into the fold later on, however, which it did. In the 1960s, with growing national attention on poverty and social problems, Congress again considered the question of where the social sciences fit within the larger picture of federal funding and administration, and this time many called for a separate National Social Science Foundation that could facilitate research into the country's domestic issues. In each instance, arguments were made for and against the alignment of the social sciences with the natural sciences, and every time the similarity or difference in their respective research methodologies was a key factor in the debate.[37]

With the political currents of the 1980s moving against federal research funding (just after the cultural backlash against social science interventions in schools incited by the MACOS project), the social scientists hardly felt secure. The fact that NSF gave only "marginal attention" to the social and behavioral sciences in its budget plan in 1986 was a red flag for the AAAS, which sought to safeguard those fields within the realm of the sciences writ large. Project 2061 accordingly laid out a broadly inclusive vision of what counted as science. In the original proposal the group highlighted its broad vision of the sciences as something that made its work distinct from the post-Sputnik reform movement. "For the most part," wrote Rutherford, the earlier curriculum projects "stayed well within disciplinary boundary lines, and paid little attention to the engineering, health, social, behavioral, and applied sciences, or to technology. Project 2061 intends to scan the entire landscape of science."[38]

Given the traditional set of school science subjects (biology, chemistry, physics, and general science), it was a bold move for Rutherford to include a panel devoted entirely to the social and behavioral sciences. Yet even with the avowed commitment to those domains, methodological issues emerged to complicate matters. In the panel report, written under the direction of Harvard psychologist Mortimer Appley, much attention was paid to what a scientific approach to the study of humans and human systems entailed. The group made clear that systematic, rigorous investigations produced outcomes far different from those that might "be drawn from the various 'common sense' observations" people rather casually tend to make.[39]

The details of these investigative approaches were laid out in a separate appendix titled "Scientific Study of Social and Behavioral Phenomena." It began with a fairly standard account of method (which included the necessity of forming hypotheses and the need to test these against data as well as alternative possibilities). But then the group delved into the problem of bias, the differing approaches used by scientists operating in various "schools of interpretation," and the challenges of studying subjects that have histories, that respond to and react to the communication of findings, and that can behave differently as a result of simply being observed. The panel described as well the difficulties involved in the definition and measurement of human and social attributes, problems of sampling, and the difficulty of accurate prediction. It was a thoughtful and accurate depiction of the research issues social and behavioral scientists typically faced.[40]

The treatment of the social sciences that appeared in the initial draft of the 2061 consensus report, however, glossed over nearly all of this methodological nuance—possibly due simply to space constraints. This bothered Appley, who voiced his objections at the June 1987 meeting of the National Council on Science and Technology Education. He took issue with the various hedges and qualifications that he found, including statements such as "while rarely conclusive," the "diversity of methods . . . may be more fragmented," and "a science-based unified view is not yet attainable." The whole thing, he felt, was "a little pejorative—that if we are not the natural sciences, we must be the unnatural sciences." He stated plainly, "Either we do have the possibility of applying scientific study to these phenomena or we do not." Another participant agreed that there seemed to be a "deep ambivalence about the rigor of the social and behavioral sciences" in the document and added, "I don't think apologies need to be made." The points were duly noted by the council co-chair William Baker, who sought to reassure Appley that the position of 2061 was "that this is and can be science."[41]

The strategy Rutherford and the council used in the final version of *Science for All Americans* was simply to assert the scientific legitimacy of the social sciences and leave out any substantive discussion of methodology. "Science can be thought of as the collection of all the different scientific fields . . . from anthropology through zoology," they noted in the first chapter, and they acknowledged that "they differ from one another in many ways." But they went on to state unequivocally: "With respect to purpose and philosophy, however, all are equally scientific and together make up the same scientific endeavor." When it came to the social and behavioral sciences chapter (chapter 7), they limited their discussion to the various concepts and key aspects of human society. The goal, they stated up front, was to sketch "a comprehensible picture of the world that is consistent with the findings of the separate disciplines within the social sciences . . . but without attempting to describe the findings themselves or the underlying methodologies."[42] The omission likely served them well.

As much as Project 2061 sought to raise a bulwark against those who would exclude the social sciences from what was seen as the field of legitimate science, it sought as well to build a wall against those who sought unauthorized entry. Here members of 2061 were concerned with those advocating the inclusion of "creation science" in school science classrooms. The United States Supreme Court had ruled in 1968 that states could no longer ban the teaching of evolution in schools in favor of teaching creationism. The appearance of this new "science" was part of a strategy on the part of religious fundamentalists to undermine the legitimacy of evolution by presenting the biblical creation story alongside it as another scientific explanation for the origin of life (humans in particular) on Earth. If creationism was science, so the argument went, then it would have a claim to be included in the science curriculum and thus could offer a counterbalance to godless evolution.[43]

The strategy of casting creation as a legitimate scientific explanation was a clear affront to the AAAS and signaled an entirely new threat to scientific authority. Ronald Reagan's statements during his presidential campaign supporting the teaching of creationism (which were soon followed by his call as president for a constitutional amendment to permit organized prayer in public schools) brought the issue even closer to home. In response, the association's board of directors passed a resolution in 1982 condemning the "forced teaching of creationist beliefs in public school science," hoping to draw public attention to the fact that "creationist groups are imposing beliefs disguised as science on teachers and students to the detriment and distortion of public education in the United States." Science, the board asserted, "is a systematic method of investigation based on continuous experimentation, observation, and measurement leading to evolving explanations of natural phenomena, explanations which are continuously open to further testing." Creation science, on the other hand, "has no scientific validity" and "should not be taught as science." The message from the AAAS was clear and unambiguous.[44]

The evolution / creation clash was a hot topic of conversation among the 2061 panel members when they started meeting in 1985. Helen Ranney, professor of medicine at the University of California, San Diego, and chair of the biology panel, bluntly pointed out: "It's a problem again." In fact, during the drafting of the group's report in the fall and winter of 1986, oral arguments in *Edwards v. Aguillard,* a case testing the constitutionality of Louisiana's balanced-treatment law, were heard before the United States Supreme Court, and the Court's decision (finding it unconstitutional) was handed down the following June, days after the meeting of the National Council. With the creationist threat front and center, George Bugliarello insisted repeatedly during these meetings that the report deal with the "separation of scientific truth from religious truth," making a point that seemed to look back to more traditional views of scientific authority. Others similarly emphasized the importance of "verification to distinguish science from error, fraud, dogma, or superstition," highlighting the view of science as "verified" knowledge as a means to distinguish it from creationism and the other pseudosciences of the time.[45]

Demarcation was clearly the goal. In his notes for the chapter on the nature of science, Rutherford wrote, "The label 'scientific' is widely used, often speciously. The effectiveness of certain foods and medications, health and beauty treatments, energy-free machines, money-making schemes, horoscopes and other fortune-telling methods are often declared by their advocates to be scientific." He went on: "So, too, do some advocates of unscientific beliefs about the origin of species, for example, . . . claim the mantle of science." What the public needed to have, in his mind, was a "questioning attitude." This, he wrote, targeting the creationists directly, "is the hallmark of people able to tell science from pseudoscience and to know that labeling something 'scientific' does not necessarily make it so."[46]

The plan was for the "nature of science" chapter to do most of the work on the demarcation front, where a careful listing of features and characteristics would make obvious where the line between science and non-science should be drawn. (This line became even more important when it was revealed in the spring of 1988 that the First Lady, Nancy Reagan, regularly relied on astrology in her daily decision making.) Rutherford needed to be careful, however. Emphasizing the need for empirical evidence, logic, and prediction all worked in science's favor. But including more recent insights from science studies about the tentative, sometimes revolutionary nature of science could be more problematic. Harvard geologist Raymond Siever pointed this out during the discussion of the draft consensus report. He worried about the report using language such as "that science teaches the truth as of now," since "that argument has been twisted by the creationists." Rutherford admitted this was a problem they struggled with throughout the report—"trying to get a reasonable balance between the notion that in science things change and they have to be accommodated—sometimes it is revolutionary and sometimes it is not—and, on the other hand, not giving the impression that nothing counts because tomorrow it is going to be different." In the case of evolution, he hoped that the report would convey to teachers that they "should not use authority to present the idea" but should "make it clear why the scientific community believes it is a fact, and not just an idea that will go away."[47]

The "nature of science" chapter in the final version of *Science for All Americans* did its job as planned. It laid out the essential elements of the scientific worldview—that the world is understandable, that scientific ideas are subject to change, but that such knowledge is also durable. Rutherford then explained that "science cannot provide complete answers to all questions," pointing out that certain beliefs and assumptions, such as "the existence of supernatural powers and beings," cannot be "usefully examined" by science, and he admitted that some science "is likely to be rejected as irrelevant by people who hold to certain beliefs (such as in miracles, fortune-telling, astrology, and superstition)."[48] Beyond this, he picked no fights with the creationists, leaving the positive statements about what science was to speak for themselves.

The chapter on the life sciences (chapter 5, "The Living Environment") similarly provided a nonconfrontational description of the current state of evolutionary understanding, following the recommendation of University of California, Davis, biologist Francisco Ayala to emphasize the "facts" of evolution and avoid at all cost the phrase "theory of evolution," lest any doubts be raised about its validity. The only place the creationists were addressed directly was in the chapter on historical perspectives in science (chapter 10), where the report acknowledged that for "some people" evolution "violates the biblical account of the special (and separate) creation of humans and all other species." They emphasized that there was no debate, however, over the question of "whether evolution occurs."[49]

Although born from the economic crisis of the late 1970s, the substance of *Science for All Americans* as it appeared as the very end of the 1980s embodied only remnants of the nation's earlier concern with American job losses, lackluster economic growth, and energy shocks (the inclusion of technology and engineering seeming to be the only nod to the original economic motivation). The report, instead, followed the interests of Rutherford and the members of the professional science establishment who pulled it together. It presented a vision of science literacy in answer to poor student test scores, declining academic achievement, and a public that was losing interest in—and respect for—the research enterprise and the expertise scientists possessed. One unfriendly critic was spot on in his summary of the publication, writing that it "has much more to do with learning *about* science than immersing oneself *in* science." This was just the sort of understanding, however, that Rutherford and the scientists felt was most needed to protect and bolster the enterprise in which they were engaged. They embraced the "science for all" approach; it was important that, in the words of one council member, they address "not just the best, but the rest of the population," so that they won't "be manipulated by persons who want to drive our society in ways that are ultimately not going to be in our interest." Put another way, as Mike Atkin of Stanford University pointed out in testimony to Congress, the country didn't need "a huge fraction of the population to take up careers in science." Resurrecting a comment made by the physicist Edward Teller in 1958, he explained that, rather, "what the United States needs . . . is science 'fans.'" In this way, Project 2061 was not all that different from the curriculum reforms of that earlier era.[50]

10 Science in the Standards Era

HALLEY'S COMET, the measure of Project 2061's progress, was only twenty years gone on its journey into the solar system—not even halfway to its farthest point from the sun—when yet another science education crisis hit the United States. This time the threat was from the newly interconnected, globalized economies of India and China. The new challenge was captured vividly by Pulitzer Prize–winning columnist Thomas Friedman in his book *The World Is Flat: A Brief History of the Twenty-First Century,* published in the spring of 2005. There Friedman described how a host of innovative technologies (such as personal computers and the internet) and business practices (offshoring, insourcing, and supply chaining) converged to upend the traditional job market. Workers in the United States suddenly had to compete for jobs with workers all over the world; people who had been geographically remote were now, with "the creation of a global fiber-optic network," all "next-door neighbors."[1] The challenge was most acute in the fields of science and technology, where knowledge workers (as they had come to be called) were competing with the hundreds of thousands of scientists and engineers being trained annually overseas in countries that produced far more such workers than the United States (and who cost far less to employ). American science education, in Friedman's view, needed to catch up, and fast.

The World Is Flat caught the country at an anxious and profoundly vulnerable moment, not long after the shock of the 9/11 terrorist attacks in 2001, when globalization as an idea suggested both indiscriminate harm as well as expanded international commerce and wealth. Playing to these anxieties, the book became a worldwide bestseller, going through three editions in the first two years after publication. Friedman's argument, one reviewer wrote, was "lively, provocative and sophisticated," and while "we've no real idea how the 21st century's history will unfold, . . . this terrifically stimulating book will certainly inspire readers to start thinking it all through." Such thinking began almost immediately in an array of congressional hearings convened to grapple with the pressing issue of America's faltering technological competitiveness. Elected representatives and experts called to testify repeatedly invoked Friedman's depiction of the challenges ahead. One witness commented that she had heard reference to this "very popular book" for the "um[p]teenth time." Friedman had clearly initiated a drumbeat call for action.[2]

Although concerns about the country's technological competitiveness had been raised prior to Friedman's book, they now coalesced into concrete policy proposals aimed at improving the science preparation of American students. The most significant of these were contained in the report *Rising above the Gathering Storm,* published by the National Research Council in the fall of 2005. In that report, the authors repeated earlier warnings that "the inadequacies of our system of research and education pose a greater threat to U.S. national security over the next quarter century than any potential conventional war we might imagine," and quoted as well the director of the National Science Foundation, who predicted that "the big winners in the increasingly fierce global scramble for supremacy will be those who develop talent, techniques and tools so advanced that there is no competition." Friedman called it a "quiet crisis." The authors of the *Gathering Storm* report did all they could to ensure it didn't stay quiet for long.[3]

The impact of this new global challenge on science education, however, was less dramatic than what had been wrought by the crises of the past (such as those triggered by Sputnik and the *Nation at Risk* report). In terms of public understanding of science, it merely accelerated a downward trend. The reform initiative set in motion by AAAS's Project 2061 was already rolling along by 2005 and had become the leading edge of the new movement for educational standards in the United States. In 1993, Rutherford and colleagues converted *Science for All Americans* into a more classroom-friendly set of benchmarks for achieving science literacy. Two years later, the National Science Education Standards (NSES) were added to the mix.

With the implementation of a standards-based science curriculum and standardized assessments, the ideal of scientific literacy for citizenship—the central goal advanced by Rutherford and others at Project 2061—was gradually slipping away. In a world of content standards, complex ideas about the nature of science and scientific inquiry were set aside in favor of content mastery. When the "flat world" and *Gathering Storm* crisis hit ten years later with its unrelenting calls for more advanced technical training, it pushed these science literacy goals further toward the margins than they had already been. More than anything, the globalization crisis succeeded in solidifying a view of science education as a means of workforce development in which teaching about the methods of science in American classrooms, for all practical purposes, had a low priority indeed.

* * *

The publication of *Science for All Americans* in 1989 completed phase I of the AAAS education reform project. Phase II aimed to "transform the content recommendations of Phase I into *educational guidelines for sustained and purposeful reform.*" This, according to Rutherford's plan, was where teachers, curriculum directors, and educational psychologists were to be brought on board. The Project 2061 team assembled working groups with personnel located in six school districts across the country. Their "enthusiastic support" was key. "Unless this [support] is achieved," Rutherford wrote, "the likelihood of a curriculum that meets the goals identified in Phase I being implemented in classrooms is minimal." Following a summer retreat in Colorado during which the new team members were tutored on the vision and goals of the project, they, along with Project 2061 staff, set to work breaking down the content of *Science for All Americans* into discrete learning outcomes arranged in a developmental progression across the various grade levels. What Rutherford had initially referred to as a "curriculum profile" was eventually published as *Benchmarks for Science Literacy.* The task of translating the nature of science and scientific inquiry as laid out in *Science for All Americans* into these benchmarks hit some unanticipated snags along the way.[4]

From the start, the leadership of Project 2061 sought to treat the methods of teaching as separate from the content to be taught. They recognized, however, that such a distinction was difficult to maintain, for in considering the science literacy outcomes they were seeking to instill in students, "essential implications for teaching arise." They subscribed to the belief, as had a long line of science educators before them, that teaching science should parallel the process of science itself. This was "the first lesson," as Rutherford put it, "the goal beyond all others." And so they modeled their pedagogical principles on the methods of scientific inquiry. First among these was that learning science "should be a lively affair—active, participatory, non-bookish." It was essential that students do things, "the same kind of things, in essence, that scientists, engineers and mathematicians do in their work." They recommended that "teaching should begin with real phenomena" just as science does and that "the collaborative nature of the scientific and technological enterprises should be strongly reinforced by frequent group activity in the classroom."[5]

These principles were presented in the final version of *Science for All Americans* in a separate chapter near the very end of the book and were thus positioned apart from the science literacy goals that preceded them; they appeared to readers as merely the means to the content ends, that is, as pedagogical methods rather than what was to be learned. As a result, the actual doing of science seemed to be missing from the final product.[6]

The translation of the ideas about the nature of science from *Science for All Americans*—which entailed the scientific worldview, inquiry, and the scientific enterprise—into the atomized benchmarks exacerbated the problem, which external reviewers were quick to point out. A group of Pennsylvania teachers commented that while the emphasis on active learning in the document was commendable, "nearly all of the benchmarks are stated in terms of things that students should 'know.'" Another set of reviewers, from a school district in Rhode Island, wrote in a similar vein: "Your existing objectives can be assessed by written tests and NOT by performance tests. We question whether students will 'do' science using your Benchmarks, or will they still be engaged in passive listening to content-based instruction?" Gerard Piel, who had been with Project 2061 from the start, may have put the matter most eloquently (and emphatically). "What I hoped to see," he wrote in his review, was a plan for the "reconstruction of the teaching-learning setting in our schools to the end that children will learn how we know what we know, acquire immunity to authority and seek the autonomy that is the essence of self-governing citizenship." For him the benchmarks, by their very nature, worked against this goal. "What I fear from 2061 now is that, for all the insistence on learning-by-doing, the goal-setting syllabus—given the testing and outcome neurosis that besets American education—

will suffocate the best-intentioned effort to set children free to be autonomous citizens. Say it ain't so!"[7]

The project leaders struggled with how to address these concerns. In a memo to Rutherford, associate director Andrew "Chick" Ahlgren assessed the situation: "We have heard a fair number of plain folk who object to Benchmarks for not being hands-on or performance based. We have heard similar objections from some very sophisticated folk, too." "Remember the long march from the 50's in which science educators have finally moved away from just knowing (i.e., memorizing) to 'processes' and 'hands-on' and other good things?" he asked. "Our regression to merely knowing must look like an apostasy, even a betrayal to many." Whether the reviewers fully understood the goals of the project or not, the feedback was "strong enough" to be considered "a real problem."[8]

The insistence on more "hands-on" experiences was particularly frustrating for Rutherford, given his intimate knowledge of the fate of the earlier, inquiry-based projects. "Hands-on and learning by doing," he explained, "are powerful catch phrases," and the issue, as he saw it, was that "hands-on has been pretty nearly sanctified, putting it beyond criticism." But it was important to ask questions such as "When is learning by doing counterproductive?," "What conditions must exist for hands-on to work?," and "How can we be sure that 'action' doesn't replace thought?" "A more serious concern," he wrote, "is the tendency to treat hands-on as an end in itself instead of as a means to an end," thus making clear his emphasis on the difference between the process of teaching and what was to be taught. "[*Science for All Americans*] and *Benchmarks* both describe desired outcomes in terms of learning goals—which is to say, as ends. That is their primary purpose," he concluded.[9]

The means/end challenge was a steep hill to climb, and it appeared wherever the Project 2061 staff turned. In other places, reviewers asked why laboratory work wasn't part of the *Benchmarks*—another unquestioned expectation of science instruction since the post-Sputnik reforms (and as far back as Edwin Hall and the Harvard list of experiments). Rutherford knew this, of course; he wrote, "No orthodoxy is more entrenched in science education than the need for laboratory experience. Yet laboratory skills are not among the specified learning outcomes in either *Science for All Americans* or the forthcoming *Benchmarks for Science Literacy*." Defending his vision of science for the citizen, he explained that "adults cannot reasonably be expected to *do* laboratory science in everyday life—unless, of course, their occupation is scientific—any more than they should be expected to compose music or try a court case, even though we would like them to appreciate music and understand the judicial process." He pointed out, though, that "Project 2061 strongly advocates that students conduct actual investigations, including some that required laboratory work." The distinction he again tried to make was that laboratory work, "properly conducted," was at best "a necessary means to an important end, namely learning how science works," not an end in itself.[10] The simple act of doing science didn't necessarily ensure that any real understanding of the scientific process would result.

Without some clear performance standards, though, some teachers couldn't imagine how to get students to understand "how science works." The "nature of science" benchmarks seemed to them to be too fuzzy to implement effectively. "We think there is a need for a tighter definition of what science is," wrote a group of New Jersey teachers. "The science is not right because too many other things have been mixed in with it." There were no methodological "steps," to be sure, and the insights from science studies about knowledge being stable but tentative must have seemed difficult to make sense of pedagogically. These concerns about clarity were shared by reviewers from one of the textbook publishers who felt that that the "Nature of Science" chapter lacks "the scope and organization to provide the guidance that is needed for instructional planning and materials development." Teachers had been conditioned to believe that anything that wasn't content knowledge must be process, and that process was usually taught through student performance. "'Process' in one form or another has been the mantra since the 1950s," Rutherford lamented, "never mind that the 'bad' things—memorization, authoritarianism—are still prevalent." He knew the dirty secret that "teachers often deal with process by having kids memorize steps and procedures—remember 'the scientific method.'" No matter how much emphasis might be placed on students doing science as a means of understanding science, it was treated as content in the end anyway.[11]

Rutherford and Ahlgren dealt with the situation as best as they could. They made reference to performances in addition to knowledge outcomes where it made sense, and in a chapter at the end titled "Issues and Language" they provided an extended justification for their use of terms such as "know" and "knows how" in place of action verbs that might capture the "observable behavior" of students. They also explained what they saw as the problem with a mindless emphasis on "hands-on" learning, noting (just as Dewey had observed decades earlier) that "it is possible to have rooms full of students doing interesting and enjoyable hands-on work that leads nowhere conceptually," and they added clarifying language on how laboratory experiences might be used to help accomplish their science literacy objectives ("reduce the number . . . and eliminate many of their mechanical, recipe-following aspects").[12]

The final version of *Benchmarks* reflected all the central elements of *Science for All Americans*—the complex (for some reviewers too complex) treatment of the nature of science that drew on the then-current science studies literature, providing criteria to distinguish science from pseudoscience; an account of scientific inquiry that distanced it from the traditional "rigid sequence of steps commonly depicted in textbooks as 'the scientific method'"; and the explicit inclusion of benchmarks for high schoolers focused on the importance of peer review and the self-governance of the scientific enterprise.[13]

•   •   •

It was in the midst of developing *Benchmarks* that the push for national standards began to sweep across the country. Following the shock of the *Nation at Risk* report in 1983, various initiatives were launched that sought to address the perceived deficiencies in American public education. Project 2061 was one direct result in the area of science. But there were others as well, aimed at bettering education on a more comprehensive scale. California, for instance, initiated a revision of the state's curricular frameworks and assessments across all subjects in the hopes of improving student achievement. More significant were the efforts of the National Governors Association, whose concerns about jobs and economic growth led to the 1989 Education Summit in Charlottesville, Virginia; the summit resulted in a unified call for national standards in education and a goal (one of six) that U.S. students "be first in the world in mathematics and science achievement" by the year 2000. With the federal government's commitment to these outcomes, the Department of Education, in collaboration with other federal agencies, such as the National Science Foundation, began awarding grants to scholarly associations for the development of standards in all the main academic school subjects. The National Research Council (NRC), the operational arm of the National Academy of Sciences, stepped in to develop those in science in 1991.

Four years later the National Science Education Standards were released to the public. These followed the science literacy emphasis of Project 2061, for the most part. One departure, however, was the explicit attention paid in the NSES to inquiry as both means and end.[14]

The role of the NRC in developing the new standards was a compromise negotiated between the two leading science education organizations of the time. On the science side was the AAAS with its Project 2061, which had nearly completed its work defining science literacy for the public and parsing those learning outcomes into the age-graded benchmarks. From the teaching side was a program initiated in 1989 by the National Science Teachers Association (NSTA) called "Scope, Sequence, and Coordination" (SS&C). SS&C was explicitly aligned with the philosophy of Project 2061, but it aimed at a much shorter timetable for implementation. The NSTA modeled its project on science education practices common in competing industrial nations, organizing interdisciplinary science instruction by grade level; that is, students would learn material from all the major science subjects every year rather than take individual subjects one by one. When the time came in 1991 for the Department of Education to choose a group to write the science standards, Bill Aldridge, the executive director of NSTA, asked the chair of the NRC to take on the task rather than either NSTA or AAAS. Encouraged by the leaders of other science and science education groups, the NRC agreed.[15]

Although the NSTA program had some influence in the local sites in which it was developed (it was piloted extensively in California schools), the work that Project 2061 had done defining science literacy for the American public since the mid-1980s set the pattern for the standards-based efforts of the 1990s. The NSES, by some accounts, adopted the Project 2061 concepts and framework nearly unchanged (although they limited their focus to the natural sciences, excluding the social sciences and mathematics from their purview). When it came to standards related to understanding what science was as an organized activity distinct from disciplinary content, the NSES used the same "nature of science" language found in *Science for All Americans*. The authors freely admitted in the text that its standards on this topic were "closely aligned" with the "nature of science and historical episodes" found in *Benchmarks*.[16]

The political concerns that drove attention to the nature of science in the Project 2061 publications—the need to protect the intellectual authority of science and shore up public support—were still present and, in the eyes of some scientists, becoming more prominent. Federal funding of science declined in inflation-adjusted dollars following the winding down of the cold war and the dissolution of the Soviet Union in 1991, and some of the key "big science" projects of the period, such as the Hubble space telescope and the Superconducting Super Collider particle accelerator (subsequently killed by Congress in 1993), had begun to attract critics who felt that scarce federal dollars could be better spent in areas more directly tied to economic growth.[17]

In the academic world, what came to be called the "science wars" broke out in the midst of the NRC standards work with the 1994 publication of the book *Higher Superstition: The Academic Left and Its Quarrels with Science.* Written by biologist Paul Gross and mathematician Norman Levitt, the book defended the authority of the natural sciences against various feminist, multicultural, and postmodern academic critics who (inspired in part by what Gross and Levitt viewed as a misreading of Kuhn's ideas about scientific change) increasingly argued against the privileged place of scientific knowledge in contemporary society and culture. These battles were largely confined to the academic set and died down by the turn of the twenty-first century. The early salvos didn't go unnoticed by the authors of the NSES, however, who in their publication listed *Higher Superstition* as one of the "references for further reading" at the end of the chapter that laid out the various perspectives informing their report.[18]

As much as the NSES followed the structure and content of Project 2061, one of the key differences was the manner in which it handled scientific inquiry. *Benchmarks* had included knowledge *about* scientific inquiry among its learning outcomes related to the nature of science.[19] But the description of how students themselves should engage in inquiry was relegated to a chapter toward the end that dealt with the pedagogical methods Rutherford and others thought would be most effective (this was the ends-means challenge they struggled with). The authors of NSES, in contrast, perhaps responding to what they had learned from feedback on *Benchmarks,* created not only science content standards but also science teaching standards. This effectively put at the front of the book the emphasis on providing learning experiences aligned with how scientists explored the natural world, something advocated by Project 2061 (and nearly all science education groups of this time), and, in the process, recast them as performance-based expectations for best science teaching practice.

Inquiry-based pedagogy was central to these teaching standards, which were the first set presented in the publication titled simply *National Science Education Standards*. Chapter 3 of this work opened with a set of foundational premises, including the idea that "what students learn is greatly influenced by how they are taught," as well as the idea that "student understanding is actively constructed through individual and social processes." These premises were elaborated in order to highlight the similarities between student learning and scientific inquiry. "In the same way that scientists develop their knowledge and understanding as they seek to answer questions about the natural world," the authors wrote, "students develop an understanding of the natural world when they are actively engaged in scientific inquiry" (an assumption shared with Project 2061).[20]

The content standards for the students came in chapter 6. There were eight in all, covering the traditional science subjects along with several less traditional topics: "unifying concepts and processes," "science in personal and social perspective," and "history and nature of science." Leading off, though, was "science as inquiry," which in the book's formulation "is a step beyond 'science as a process,' in which students learn skills such as observation, inference, and experimentation." "The new vision," it continued, "includes the 'process of science' and requires that students combine process and scientific knowledge as they use scientific reasoning and critical thinking to develop their understandings of science." At the high school level, students were expected to be able to conduct scientific inquiries—that is, "actively participate in scientific investigations"—and they were expected to develop an understanding about what scientific inquiry was and entailed. Both objectives were broken down further into more descriptive standards. The resultant learning outcomes were nearly the same as those put forward in *Benchmarks*. But the goal of having students be able to conduct inquiries on their own provided the performance standards that the *Benchmarks* reviewers had felt were missing.[21]

The continued commitment in the NSES to doing science as a means of learning science had been bolstered over the years by research in a number of fields. Studies in the area of cognitive science, which emerged in the 1950s under the leadership of the Harvard psychologist Jerome Bruner (a key player in developing the MACOS curriculum), began to examine the mental processes, thought structures, and creative insights involved in human reasoning. This work, which was a reaction against the dominant research paradigm of behaviorist psychology, was largely based on a belief (not unlike that advanced by Dewey years earlier) that scientific thinking provided the best model for rational thought generally. Education researchers soon paired the advances in cognitive psychology with the emerging understandings of the social and epistemic elements of scientific process contributed by science studies scholars (beginning with Kuhn in the 1960s) to lay the foundation for new thinking about student learning in the 1990s, which was described in the 1999 landmark NRC report *How People Learn*.[22]

The conception of inquiry advanced in the NSES (ideas about the psychology of student learning from the 1990s notwithstanding), though, was really a blend of the old and older. Details of the NRC vision were spelled out in a follow-up publication, *Inquiry and the National Science Education Standards*, published in 2000, which was designed to be a practical guide for classroom implementation of the NSES recommendations. In this guide the authors reiterated the familiar parallels between scientific inquiry and student learning and then, in a chapter outlining the historical and philosophical background of inquiry teaching, described the contributions of Dewey (specifically his call for more attention to scientific process from his 1909 AAAS address in Boston alongside the steps in his "complete act of thought" from *How We Think*) and Schwab (highlighting his elaboration of the two modes of inquiry teaching—teaching science *by* inquiry and teaching about science *as a process of* inquiry). The ideas of these earlier curriculum reformers were highly influential in shaping the NSES vision. They were, the *Inquiry* authors noted, "of special significance to this volume" in that they "widely disseminated the idea of helping students to develop the skills of inquiry and an understanding of science as inquiry." The National Science Education Standards were fully committed to these dual learning outcomes. Their resurrection, though, seemed to pass over without comment the documented failure of inquiry teaching in the lives of earlier reforms.[23]

•   •   •

Throughout most of the 1990s, the National Science Foundation spent $25 million per year on the development of standards and standards-based curriculum materials for math and science education, with the focus in science tilted heavily toward goals related to inquiry and the nature of science. Despite this well-funded push for reform, teachers struggled mightily in their attempts to enact the vision put forward by either Project 2061 or the National

Science Education Standards (which was undoubtedly no surprise to Rutherford). Angelo Collins, the director of the NRC project, identified what she saw as the most likely barriers to implementation at the time. Foremost among these was the lack of a shared vision of the goal. Conflicting views of science education, she wrote, were present at all levels: among students who "believe that science is a difficult subject" meant only for "selected students" rather than simply "a process of asking and answering questions"; among teachers who saw science as a "rhetoric of conclusions" (a favorite expression of Schwab's) rather than "as an adventure"; and among textbook publishers and policymakers who might fail to appreciate the depth of changes necessary and the time needed to make it all work. Nearly all these factors did indeed frustrate the implementation of the standards and thus compromised the goal of a science-literate citizenry as envisioned by the reformers.[24]

Trimming back the ever-expanding content of science to make room for understanding its methods proved to be almost impossible. The view of science as a collection of facts, concepts, and theories—as an "information subject," as it was called in the nineteenth century—had a strong hold on the minds of school district leaders, parents, and students. One external evaluator found that, even with concerted efforts to get the word out about the new vision of science education, "few teachers agreed" with the reform mantra that "less is more." Even the authors of the standards themselves found it difficult to follow that advice. An outside evaluation commissioned by the NRC on the impact of the standards and published in 2003 found that "the sheer number of topics in the NSES exerts pressure simply to 'cover' the content." Another study (published a year later, eight years after the release of the standards) found that "only 21 percent of science lessons nationally"—just over one in five—provided "experiences for students that clearly depict science as investigative in nature." At the high school level, the numbers were even less encouraging. Only 14 percent of the lessons observed were deemed to have a positive effect on student abilities to engage in scientific inquiry.[25]

Science textbooks, which teachers typically relied on in planning their lessons, were of little help. Researchers looking for the influence of the reforms in biology, for example, found that publishers "simply added more material to what was already present" in order to align with the standards. The result was books that were "just too large, still too encyclopedic." However much they might have wished to support the reform effort, publishers argued that their hands were tied by the marketplace. "Even though people ask for 'less is more,'" one commented, "when they make their decision [about what to purchase], they want everything . . . to cut content would be financial suicide." The seemingly ever-present need to cover the content of science made teaching about process a low priority. "They are still trying to teach everything," observed one national education leader. "If we really take inquiry seriously, and it takes a month or two or three, what are we giving up?" It was obvious that they weren't really taking inquiry seriously at all—that's what they were giving up.[26]

For those few who did make an effort to adopt inquiry teaching, the results were often disappointing. Studies found that teachers often made only piecemeal changes in their teaching, typically at a surface level, and not in ways that reflected the robust vision of inquiry advanced in the standards. "'What we have is teachers using hands-on [lessons], using cooperative learning' at the expense of 'teaching for understanding,'" one evaluator lamented. This—treating hands-on activity as an end in itself—was just what Rutherford had feared. There was a tendency as well among pre-service teachers (and likely in-service teachers) to rely on their own personal "folk theories" of inquiry, which often equated inquiry with "the scientific method," the bête noire of reformers since the 1950s. This perhaps wasn't surprising given the cultural pervasiveness of the idea and the fact that teachers continued to see the method legitimized through its regular appearance in teacher journals or its brief but iconic treatment in the first chapter of many science textbooks.[27]

Beyond textbooks and their own personal conceptions of how science worked, teachers had little else to go on. There were scant laboratory materials to help teachers create meaningful inquiry experiences for students or even to help them see what science as inquiry looked like. "The 'vast majority' of materials being used by teachers" fell short and were "not in line with the NSES," evaluators concluded. (The troubles BSCS had had with scientific supply companies in the 1960s appeared to be an enduring obstacle to reform.) The following year the NRC's Board on Science Education, chaired by University of Colorado Nobel laureate Carl Wieman, assembled a committee to examine specifically the role of the laboratory in high school teaching. Its findings, released as *America's Lab Report* in 2005, found what many would have easily guessed—that student laboratory activities were

often ill-defined and poorly integrated into other aspects of classroom instruction. They typically consisted of a set of step-by-step instructions students followed in hopes of arriving at some predetermined result. The high
school labs of the early twenty-first century were, in this way, not significantly different from those developed by Edwin Hall or Charles Bessey in
the 1880s and 1890s, asking students only to engage in the mindless manipulation of apparatus with little consideration of how it related to the
formation of scientific knowledge.[28]

One lever of change that might have been used to move science education toward inquiry-based teaching was assessment, particularly given the
growing dominance of standardized testing and accountability systems at
the state level. It was well known that, despite nearly everything else, teachers
taught to the test. This was especially true if some significant consequences
were attached to the results. But tests that might actually measure student
understanding of the nature of science or scientific inquiry (let alone the
ability to competently engage in scientific work) never materialized in the
years following the launch of the standards. The formal assessments common
in this era focused predominantly, and perhaps not surprisingly, on the facts
of science. "So much of what is in the standards," observed one state education consultant, "is not adequately measured by traditional multiple-choice
norm referenced tests." But these were the kinds of tests that were easiest
(and least expensive) for states to develop. Testing a student's knowledge of
the parts of a cell was simply more straightforward than, say, testing a student's ability to evaluate ambiguous data from an experiment or to assess
the validity of a knowledge claim. If an exam "only asks questions about
what does this mean, fact, fact, fact, fact, and doesn't talk about process,"
worried one science educator, "then, of course, teachers are going to . . . say
'forget these ways of teaching,' which talk about process, which talk about
experiments." The "testing and outcome neurosis" Gerard Piel had seen
spreading across the country in the early 1990s seemed to have had the suf-
focating effect he had feared.[29]

Perhaps the most damning statement to be found was in the 2003 NRC
investigation of the impact of the NSES. After the investment of tens of millions of dollars and the deliberate efforts of scores of scientists, education
researchers, school leaders, and policymakers, the authors concluded that
"little has actually changed" when it came to everyday teaching in classrooms. It was, for all practical purposes, the same conclusion reached in the

1977 evaluation of the first wave of inquiry-focused reforms after Sputnik. Having teachers engage students in inquiry as a means to learn about science simply didn't work at scale in American classrooms. Some raised questions as to whether it really worked at all. A team of researchers in 2006 faulted this abiding faith in inquiry-based pedagogy, correctly tracing it to the curriculum projects of the late 1950s. They wrote that it was at that point that "educators shifted away from teaching a discipline as a body of knowledge toward the assumption that knowledge can best or only be learned through experience that is based only on the procedures of the discipline." This, they suggested in an understatement, "may be an error," for "it appears that there is no body of research supporting the technique."[30]

*   *   *

In 2004, the National Science Board (NSB) released its biennial assessment of the state of science and engineering research in the United States, *Science and Engineering Indicators,* which tracked the usual markers of science health, such as the amount of research funding, number of patents awarded, research articles published, degrees awarded by academic major, and so on. In its discussion of secondary education, the report told a familiar story of flat student performance on national assessments, a persistent achievement gap between white students and racial and ethnic minorities, and U.S. students falling behind their international peers. It noted as well that the efforts begun in the 1990s to define and implement educational standards, such as those included in *Benchmarks for Science Literacy* and the NSES, as a means to improve student achievement (particularly through their linkage with state standardized assessments) had produced no easily discernable effect.[31] It was a story of stagnation more than anything else.

The portion of the report devoted to public understanding of science was perhaps even more discouraging. Science content knowledge among Americans was not improving, and when it came to knowledge about science and how it worked (the key goals of science literacy), the report noted, "most Americans [two-thirds] do not clearly understand the scientific process"; moreover, "belief in various forms of pseudoscience is common." Such findings were not new; indeed, they had become all too familiar in the NSB reports over the years. What was unusual, however, was the inclusion with the *Indicators* of a first-ever companion report, this one titled *An Emerging and Critical Problem of the Science and Engineering Labor Force.* This workforce-focused addendum signaled a new direction in science education policy. Although the inquiry approach to science education had not proven to be effective, this shift in thinking about the overall purpose of science education essentially abandoned any effort to address the ongoing public misunderstandings of the intellectual and institutional workings of science in the United States in favor of workplace preparation.[32]

The reason for the policy shift can be traced to the terrorist attacks of September 11, 2001. In the aftermath of those tragic events, the U.S. government took a number of measures to bolster American security. Within the realm of science, this included dramatically increasing the screening of visa applications from foreign scientists and graduate students in an attempt to limit the potential transmission of sensitive technical information to countries or groups that might pose a security risk to the United States. It also tightened restrictions on the communication of what it deemed "sensitive" information, a move that stirred ghosts of the Red scare from the postwar period. This drew the ire of the American Civil Liberties Union, which called the action an "assault on scientific and academic freedom."[33]

While the restrictions on communication were damaging, more concerning was the visa screening program, which had the effect of sharply curtailing the scientific talent pool available for American research efforts. Since the early 1980s, the number of U.S.-born scientists and engineers had been steadily declining. These declines had been mitigated by the growing number of foreign-born scientists who had moved to the United States to meet the needs of university laboratories. The post-9/11 security restrictions, however, had within a few short years created a scientific labor crisis of significant proportions. This sharply focused the nation's attention on what might be done immediately to increase the production of home-grown technical talent.[34]

The publication of *The World Is Flat* in 2005 highlighted in vivid detail the pressing need for knowledge workers in the new global economy. This along with the rapidly emerging shortage of U.S.-born scientists and engineers prompted calls for action among the American scientific and business elite. William Wulf, president of the National Academy of Engineering, penned an article titled "A Disturbing Mosaic," which laid out the contours of the crisis. What had prompted him to write it, he explained, was a "few lines" in Friedman's book about whether the country really had been "investing in our future and preparing our children the way we need to for the race ahead." A version of Wulf's article served as the opening chapter for *Rising above the Gathering Storm,* the bipartisan report that detailed in stark terms the threat facing the United States. Its authors relied on the conventional wisdom that "knowledge acquired and applied by scientists and engineers provides the tools and systems that characterize modern culture and the raw materials for economic growth and well being." Science education was now almost wholly viewed in light of its ability to contribute to the development of those raw materials. Headlines and op-ed pieces suddenly warned of "an erosion of the U.S. competitive edge in science," insisted that "American science is in hot water," and implored Congress to "act urgently to preserve America's edge."[35]

Prompted to action by the force of both Friedman's *World Is Flat* and the *Gathering Storm* report, Congress held an array of hearings starting in 2005 on various aspects of science education and technological preparedness. The following year President George W. Bush announced a new American Competitiveness Initiative that focused on high-tech jobs and better science and math education to prepare the public for the new knowledge economy. "You can't pick up a newspaper or magazine these days without reading about global competitiveness," noted Secretary of Education Margaret Spellings in her testimony to Congress. "We must encourage more students to take more advanced math and science classes. Employers today need workers with pocket protector skills." Others talked about getting more students into the science and engineering pipeline to replace the scientific and technical workers being shut out of the country by the new visa restrictions. "There are simply not enough students going through the K-12 system and the higher education system that are interested in science," commented California congressman Howard McKeon. House Science Committee chair Sherwood Boehlert, from New York, cast the situation in grave terms: "Every single one of us believes that K-12 science and math education is the ultimate key to our future prosperity and strength, and one might say survival as a nation. This is, indeed, a national security issue of the highest magnitude."[36]

In August 2007 President Bush signed the America COMPETES Act into law. The parts of the legislation for enhancing science education closely followed the recommendations of the *Gathering Storm* report. Among the provisions of the law were new programs to increase the number of science, technology, engineering, and mathematics (STEM) teachers through the development of new baccalaureate and master's degree programs as well as additional scholarships for students in STEM teacher education programs; to dramatically increase the number of teachers serving high-need schools who are qualified to teach math and science Advanced Placement (AP) and International Baccalaureate (IB) courses and increase the number of students in those schools taking those courses; and to fund mathematics and science education partnerships between schools and universities to improve STEM instruction. The goal, as expressed by the *Gathering Storm* authors, was to recruit 10,000 new highly qualified teachers every year. The new law targeting the AP and IB programs called for increasing by 70,000 the number of teachers qualified to teach those courses and by 700,000 the number of students successfully completing them.[37]

The passage of this legislation demonstrated a strong commitment to science education (though a number of the provisions weren't actually funded at the levels requested and some not at all). One would look long and hard, however, to find any evidence of significant attention to teaching about the process and methods of science in all of this. References to science literacy and its focus on understanding the nature of science and the scientific enterprise—the predominant learning goal of both Project 2061 and the NSES—were nowhere to be found. The emphasis was squarely on technical proficiency and science content mastery. The new teacher education programs were designed to encourage "undergraduate students *already pursuing STEM degrees* to concurrently pursue teacher education," and the new master's degree programs were aimed at enhancing the science content knowledge of existing teachers. The belief seemed to be that what teachers needed was more content expertise so that they, in turn, could funnel more students into the science career pipeline. The provision for AP and IB coursework had the same effect of moving students toward scientific and technical career paths and steering them away from learning experiences related to scientific methods or processes. AP courses in the sciences, given their role in preparing students for an end-of-year exam (which had the potential to count for college credit), were commonly faulted for their excessive emphasis on test-driven mastery of facts to the detriment of time spent in laboratory or inquiry-based work.[38]

The America COMPETES Act of 2007 marked the culmination of the nearly quarter-century push for more and better science instruction sparked by the *Nation at Risk* report in 1983. The international threat to the country's economic prosperity, first from Japan and then from China and India, provided the stimulus for action. But that same threat increasingly determined the nature of the science teaching that resulted.

The early response from AAAS's Project 2061 sought to resurrect the general education vision of the Harvard Redbook, a vision centered on an appreciation of the nature of scientific work—both as an intellectual activity and in its institutional relationship with society at large—in hopes of educating a public that would understand science as a cultural endeavor central to human progress. But Rutherford's goal for students to grasp what science was and how it worked, either through the principles laid out in *Science for All Americans* or as those principles were reconceived in the inquiry-focused National Science Education Standards, proved to be too difficult to implement. The weight of traditional ways of seeing science as information about the world and science teaching as simply conveying that information trumped the reform efforts. Moreover, systems of standardized assessments in the 1990s pushed science education even further toward facts and concepts and away from methods and process, the central focus of the reformers' science literacy ideal. The subsequent pressures from globalization and the post-9/11 science and engineering workforce crisis appeared to mark a new phase of science education in the United States, one dominated by career training and the metaphor of the pipeline, rather than a broad vision of science for all Americans.

Conclusion

THE EMPHASIS ON WORKFORCE training that came with the passage of the America COMPETES Act in 2007 didn't eliminate entirely the prospects for teaching about the process of science. The desire to instill in every generation an understanding of the scientific method seems to know no bounds, and just as there have always been science educators throughout history who have insisted on the value of learning about the methods of science, we continue to see such advocacy to this day. The most recent expression of this desire can be found in the Next Generation Science Standards (NGSS), a version 2.0 attempt to upgrade science teaching through standards published by a committee of the National Research Council in 2012. This new iteration, its drafters say, comes in response to advances in science as well as in our knowledge about education: "Not only has science progressed, but the education community has learned important lessons from 10 years of implementing standards-based education." "There is a new and growing body of research on learning and teaching in science," they continue, "that can inform a revision of the standards and revitalize science education."[1] Science education—perhaps all education—seems to be perpetually in need of revitalization. But have we really learned anything new that might help us break through and successfully teach the general public about the process of science?

Using the themes that have emerged from the historical narrative of this book offers some answers to that question. The NGSS, as currently framed, represents a continuation of the ideas that were originally put forward in the science education standards of the 1990s. In this case, though, there appears to be even more emphasis on the importance of workforce training and economic development. Nearly all the arguments made on "The Need for Standards" page of the NGSS website are grounded in this rationale. Scattered across the handful of paragraphs are references to, among other things, careers, innovation, high-tech exports, patents, and the global economy. The economic justification for science education that the Carnegie Corporation's Alden Dunham shared with Rutherford in 1984 seems right at home alongside the arguments for the current initiative. The NGSS, in fact, was itself initially funded by Carnegie, and its authors point to a key 2009 Carnegie publication, *The Opportunity Equation*—a report arguing for science education as a means to meet global competitive challenges—as a motivating factor in its development.[2]

The interest in teaching about scientific methods among NGSS advocates, as with nearly all the reformers who came before, is strong. The "lessons" that appear to have been learned about how best to accomplish this center on matters of clarity and specification. Science educators appear to have finally put to rest the notion that some holistic inquiry-based approach to teaching can actually work in practice at scale, no matter how many examples or cases might be offered for teachers to model. There was always just too much ambiguity surrounding the notion of inquiry for teachers to really know what to do with students in their classrooms.

The alternative presented in the NGSS draws on more recent work in the field of science studies following the turn toward "scientific practices." Practice serves as one of the three main dimensions of effective science education (the other two being cross-cutting concepts and disciplinary core ideas). And, in an effort to provide more explicit direction for teachers to improve classroom instruction, current reformers have broken scientific practice down into eight components that include things such as asking questions, developing and using models, constructing explanations, and engaging in argument from evidence. With this approach, the NGSS continues in the footsteps of earlier efforts, recommitting to the belief that the best model for pedagogy is the scientific process itself. "Students," they assert, "cannot comprehend scientific practices, nor fully appreciate the nature of scientific knowledge itself, without directly experiencing those practices for themselves." The sentiment expressed here, as we have seen, has deep historical roots. One can almost hear Bentley Glass—or even Edwin Hall—speaking those same words. All that's changed is the language used to characterize the process of science in any given era.[3]

The parsing of scientific practice into the eight components is a natural reaction to the problems encountered in the previous attempts to teach through inquiry. NGSS authors admitted that these past efforts were "hampered by the lack of a commonly accepted definition of [scientific practice's] constituent elements."[4] But if history is any guide, it won't matter how carefully packaged this new version of the scientific process is, for if fundamental changes aren't made to how we prepare teachers and what we value as the goals of science education in the United States, the NGSS will almost surely face the same fate as the laboratory method, the scientific method, science as inquiry, and all the other variants of scientific process that came before. The eight practices, like the five steps of the scientific method from the mid-twentieth century, inevitably will be either taught in some rote fashion or neglected entirely in favor of covering the ever-growing mass of scientific content knowledge, the best efforts of NGSS proponents to warn against any disconnected treatments notwithstanding.

The NGSS, finally, provides a fitting bookend to the history of teaching science in American schools, one that highlights the changing social purposes of science education. The value of any long view of past educational practice lies in the historical trends that such a view enables us to see more clearly. The history of education reforms has often been likened to a pendulum swinging back and forth, with a focus on disciplinary knowledge or student interest, for example, coming into and going out of favor from one period to the next. As misleading as the pendulum metaphor can be, there is some truth to it here in that reforms initiated by practicing scientists in the 1880s and 1950s have indeed seemed to alternate with initiatives launched by educators prior to the 1950s and in the 1980s and 1990s.[5] But there is a more significant shift worth noting, and that is the shift from seeing the teaching of scientific methods as being of benefit to students as individuals and citizens to seeing it as being primarily of benefit to science itself as an enterprise or institution. The tipping point for that shift was World War II, when the government recognized the instrumental power science possessed and deemed it an essential element of the country's national security (initially) and economic security (later on).

From the beginning of science education in American high schools, the arguments made for its inclusion were all about what understanding the process of science had to offer the *individual*. As an information subject in Herbert Spencer's era, science promised facts about the world that would be useful to people in their daily activities. The shift to an emphasis on process as it might be learned through laboratory exercises, such as those developed by Edwin Hall or Charles Bessey, similarly offered something of value to the individual student—better observational skills, increased powers of mental acuity, or moral uplift. Such outcomes promised to provide the personal disposition that, many believed, citizens would require if the young American republic was to succeed. Although the experience of students changed with the move away from the laboratory to the psychological characterization of method that came with John Dewey and the progressive reforms in the first part of the twentieth century, the benefits still flowed from science to the student. This time, though, what carried over was a method of thinking that could be used instrumentally to solve personal and social problems. Here too the needs of the individual citizen were held paramount. Concerns about science as an institution were deemed secondary, deliberately marginalized even as the universal method of the educators was advanced at the expense of disciplinary knowledge and practice, particularly in the case of general science.

Such arguments for personal benefit or improvement of the citizenry more broadly made by scientists and science educators alike were perhaps necessary in an era during which science had to fight its way into the curriculum alongside other well-established subjects. Science education had to prove its worth somehow, whether by providing something entirely new and different (in the beginning, this was utilitarian value), by offering at least the same benefits as those provided by the other subjects in Charles Eliot's "magic circle of the liberal arts" (mental discipline and personal virtue), or by delivering singular problem-solving power (as the progressives claimed for the scientific method).[6]

But all this shifted after the war, an event Jerrold Zacharias of PSSC saw as a watershed moment. The war, he said, "changed the nature of what it [meant] to do science and radically altered the relationship between science and government." He might have added that it radically altered the nature of science education too, because after the war nearly every effort to teach students about how science worked—to reshape the public's view of what science was—centered on conveying understanding in the service of fostering respect for, appreciation of, or deference to the research enterprise. The importance of teaching about scientific methods no longer lay in how a student might use these methods to solve some personal problem (although there was hand waving in this direction at times); rather, teaching about science as inquiry was all about ensuring continued public support for scientific work. And this entailed communicating to students that there was no disembodied method that existed outside of the community of scientists and the disciplinary knowledge they used. The argument, as Edward Teller put it, was not for more scientists but for more "science fans."[7]

This changed somewhat in the standards era with the greater focus on economic development to meet the global challenge posed by competitors from Japan, China, and India. But even in this era, science education was still largely controlled by science partisans, and so the emphasis remained on teaching about the nature of science (or about science as inquiry) in order to control what counted as science and to safeguard the well-being of the broader scientific enterprise. In the postwar period, then, science education was all about ensuring that the benefits accrue first and foremost to science itself.

In the most recent period, as I've tried to show, the emphasis on science education has been justified almost solely in terms of economic development, and the result is that teaching about scientific methods has been overshadowed by an emphasis on content or, in the case of workforce training, technical skill. One might argue that this type of career preparation is an exception to the larger historical shift I've described, in that it provides an example of a benefit from science education to the individual student. But any such individual benefit is clearly overwhelmed by the advantages designed to flow predominantly to science in the interests of economic growth. Career preparation seems little more than a rhetorical crumb thrown by policymakers to counter the larger economic narrative, and it seems a disingenuous crumb at that, given that the real demand for scientists and engineers would employ only a small fraction of the number of students who graduate from high school each year.[8]

What this history shows us is the incredible malleability of school portrayals of the process of science. The manner in which this process has been presented to students—sometimes grounded in logic, other times in psychology; as an autonomous intellectual process, or as something tied closely to the tacit knowledge of a community of practitioners; and so on—has always depended on the larger purposes such portrayals were designed to accomplish, whether to empower individuals in their daily lives or to safeguard professionals from external interference.

How we taught science in each period, in other words, was not so much about how science worked as much as it was about what educators and / or scientists believed they could change about society. Each representation was designed to accomplish a particular set of social or political objectives. The take home message here is that it clearly matters how we teach science, and understanding this means, first and foremost, that we need to ask: To what end do we educate our children? It's not enough to simply attempt to convey some "authentic" account of what scientists do. As we have seen, there doesn't appear to be a way to determine what an authentic representation of scientific practice is apart from the societal purpose we intend that representation to serve.[9]

•   •   •

The history of science teaching offered in these pages isn't meant to pass any judgment on past practices. It does, however, raise interesting questions that stem from the prewar / postwar shift that I hope will stimulate discussion about the way we think about what we do in schools. The emphasis that emerged after the war (in the form of science as inquiry) came at a time of peak cultural authority for science, but also at a time of perceived public misunderstanding and suspicion. As the postwar science education reformers ramped up their efforts to clarify for the public how science "really" worked through new kinds of high school instruction, public support waned and hostility grew. Subsequent efforts to shore up science in the public mind also failed to have the desired effect, and even today, in the midst of the new focus on scientific practices (which supposedly has taken advantage of all the lessons from the past), science is under siege as never before, openly questioned, regularly neglected in policy debates, and routinely dismissed by many in the cultural arena.[10] Surely there are larger forces at work, beyond the educational system, that have brought us to this point. But the current state of affairs makes one wonder if the move after World War II toward teaching science for the sake of science, characterizing it as something distinct and separate from our everyday ways of living, is up to the task.

What we need, it seems, is a middle way between science teaching only for science or economic growth and science teaching for the individual. The ideal approach would seek to ensure public respect for scientific institutions and at the same time connect science to everyday human needs. The current NGSS initiative has a heavy focus on the practices of science as the means of learning, and even if its standards could be faithfully implemented, perhaps it is pushing things too far toward *doing* science over *understanding* science. Students are expected to engage in sense-making, construct arguments, justify their claims with evidence, develop models, and so on, and without question these are all desirable skills. The danger, though, is that the elements of scientific practice become reduced to pedagogical techniques, thus rendering the work of science and the scientific enterprise essentially invisible. Such an approach does little to help students understand the more complicated intellectual, social, and political ways that various communities of scientists go about their work in a modern democratic society.

So what, then, should students learn about how science works? If one were to hazard an alternative to the current NGSS initiative, it would begin with a clear understanding of just what the social purpose of science education should be. Since the advent of laboratory teaching in the late 1800s, efforts to teach students the methods of science have nearly always focused on some form of the citizenship goal, whether to instill the individual virtue necessary for a functioning republic or to give people the skills needed to solve problems in their everyday lives. The need for an informed citizenry is just as important today. In fact, in the current era, in which science seems to be under attack, the most pressing goal seems to be managing the relationship between our democratic political system and the community of expertise that the institutions of science represent. The public needs to understand the power and value that science offers for informing policy decisions related to how we live in the world. Citizens should be taught about the process of science in a way that enables them to acknowledge that the scientific community possesses a level of expertise that is essential if we are to make intelligent collective decisions. We should teach, in other words, that scientists know what they are talking about when it comes to things such as climate change, groundwater contamination, vaccinations, and so on.[11]

Teaching to convey the legitimacy of science in matters of fact, however, would not be enough in and of itself. Although fostering public respect for the authority of experts is clearly needed in these times, there are countless historical episodes that should warn us against moving too far in that direction lest we invite a form of scientism that could lead to government by technocracy. An ideal education in science would convey equally the essential role of the public—its political role—in steering science toward problems of pressing social need. Public support of science, in the form of tax dollars, entails some agreement that there are indeed strings attached, that scientific research should be undertaken in response to the needs of the non-expert citizens who are funding that work and that researchers should be accountable to those citizens. We need to teach not that scientific truths should be viewed as some sort of common good in and of themselves (despite what some members of the scientific community might think) but rather that their good comes from the instrumental power they provide for bettering the human condition. Scientific knowledge is a tool like any other and should be used to help us achieve our collectively determined goals. Certain truths, in other words, are more useful than others, and it's those that are worth public funding.[12]

The challenge for science education is to help students understand the point at which scientific authority bumps up against the will of the people. Certain decisions, assertions, or conclusions within a particular field of study are clearly matters for scientists to decide among themselves. Those outside the relevant community of practice simply do not have the expertise to weigh in on those conversations. But how we understand and decide to act on what scientists come to know about the natural world—and what knowledge they should seek—must be determined by the people in some way, as part of an ongoing process of political deliberation. There is, of course, no easily determined point where one can say scientific authority ends and the exercise of public preference begins. It's far more complicated than that in the messiness of real science and politics. As an educational goal, however, this seems to be a promising place to start. It could be the point at which we begin to recover some of the legitimacy of science while maintaining—perhaps even extending—the power of a politically engaged public, the middle way between science for science and science for the citizen.

Taking this as the purpose of science education would combine the visions of two figures from the history of teaching the scientific method. It would elevate Schwab's concern about ensuring the ongoing recognition of the power of scientific inquiry (in all its various forms), but it would avoid accepting unconditionally his desire to secure public funding of science without public interference. At the same time it would embrace the instrumental view of science advocated by Dewey, in which science is an intellectual tool whose value is found in its ability to address human needs, but it would avoid accepting the routinized portrayal of method that is often attributed to him.

Accomplishing this goal, it seems to me, would require a rethinking of the long-standing commitment to equating pedagogy with process—teaching the methods of science by having students do science. That approach isn't completely without merit, in that it gives students an appreciation for what it means to reason from evidence and arrive at some conclusion about a piece of the world. Too often, though, we think of the scientific process, and teach it, as a general process (or set of practices). But science isn't done in any single way, and this seems to confuse people raised with the view that science works via a step-by-step method or that an easily arranged experiment can prove what is true. What's needed in science education is to survey all the methodologies of science—experimental, historical, comparative, statistical, and so on—so that students can begin to appreciate the wide variety of intellectual work, all of it scientific, that leads to knowledge. This approach would ask for teaching that steps outside of simply doing science and instead thinks about the ways that science is done. Moreover, in order to get at the relationship between science and its role in society, we would need to add in study of the institutional context of science. Students should learn who pays for scientific research, how decisions about research funding are made and how they might be made in ways tied more directly to public concerns, and how scientific knowledge might be relied on more routinely in making policy decisions.[13]

We might very well take a page from Conant's program of science education as a means to this end, using case studies of scientific work from history or contemporary practice to understand the methods of science, and extending those case studies to include the enterprise of science itself. This latter aim is not all that different from what Rutherford envisioned in Project 2061 but never really accomplished. Such efforts might even enlist the help of social studies teachers to better address the issues related to the civic side of the question.

The continued focus on students mastering scientific practice—doing science—in the hopes that some of these larger, contextual understandings will come along for free seems misguided. Perhaps the last word on this is best left to Schwab. During the golden age of science curriculum reform, he warned that many science classrooms were "being converted into research microcosms in which every high school student, regardless of interest and competence, is supposed to act, on a small scale, like a scientist." This, unfortunately, seems to be where the current emphasis lies as well (when it isn't on the technical content of science itself). And Schwab noted too that such an approach was poorly suited to accomplishing a full understanding of the nature of scientific work. Given the choice between teaching about scientific inquiry and having students engage in the actual process of inquiry, "it is the former which should be given first priority," as Schwab said. Understanding what science is and how it works in the social context of our time is the necessary end for which we need to strive if, in Schwab's words, "we are to develop the informed public which our national need urgently demands."[14]

NOTES

INTRODUCTION

1. See, for example, Simon Newcomb, "Abstract Science in America, 1776–1876," *North American Review* 122 (Jan. 1876): 122–123; John Dewey, "Progress," *International Journal of Ethics* 26 (1916): 311–322; and "Dr. Pauling Asks Science in Politics," *Los Angeles Times,* June 14, 1954, 27. On the change in public discourse, see Daniel P. Thurs, "Myth 26: That the Scientific Method Accurately Reflects What Scientists Actually Do," in *Newton's Apple and Other Myths about Science,* ed. Ronald L. Numbers and Kostas Kampourakis (Cambridge, MA: Harvard University Press, 2015), 210–218, and Daniel P. Thurs, *Science Talk: Changing Notions of Science in American Popular Culture* (New Brunswick, NJ: Rutgers University Press, 2007).

2. Quotation from National Research Council, *A Framework for K–12 Science Education: Practices, Crosscutting Concepts, and Core Ideas* (Washington, DC: National Academies Press, 2012), 26. On the prevalence of the scientific method in schools, see Mark Windschitl, Jessica Thompson, and Melissa Braaten, "Beyond the Scientific Method: Model-Based Inquiry as a New Paradigm of Preference for School Science Investigations," *Science Education* 92 (2008): 941–967.

3. Frederick Suppe, "The Structure of a Scientific Paper," *Philosophy of Science* 65 (1998): 381–405; Jutta Schickore, "Doing Science, Writing Science," *Philosophy of Science* 75 (2008): 323–343. On where and why scientific methods are talked about, see Laurens Laudan, "Theories of Scientific Method from Plato to Mach," *History of Science* 7 (1968): 1–38; Thomas F. Gieryn, *Cultural Boundaries of Science: Credibility on the Line* (Chicago: University of Chicago Press, 1999), 37–64; and Richard R. Yeo, "Scientific Method and the Rhetoric of Science in Britain, 1830–1917," in *The Politics and Rhetoric of Scientific Method: Historical Studies,* ed. John A.

Schuster and Richard R. Yeo (Dordrecht: D. Reidel, 1986), 259–297. For some of the disagreements on method within the scientific community, see Laudan, "Theories of Scientific Method," 31–32, and Jack Morrell and Arnold Thackray, *Gentlemen of Science: Early Years of the British Association for the Advancement of Science* (London: Oxford University Press, 1981), 269–271. On the indifference of scientists to theories of truth, see Steven Shapin, "Rarely Pure and Never Simple: Talking about Truth," *Configurations* 7 (1999): 1–14.

4. Gieryn, *Cultural Boundaries of Science,* 1–25, 37–64, 65–114.

5. On the complex relationship between scientific expertise and the public, see Simon Joss and John Durant, eds., *Public Participation in Science: The Role of Consensus Conferences in Europe* (London: Science Museum, 1995); Alan Irwin and Brian Wynne, eds., *Misunderstanding Science? The Public Reconstruction of Science and Technology* (New York: Cambridge University Press, 1996); Frank Fischer, *Democracy and Expertise: Reorienting Policy Inquiry* (New York: Oxford University Press, 2009); Elizabeth Anderson, "Democracy, Public Policy, and Lay Assessments of Scientific Testimony," *Episteme* 8 (2011): 144–164; and Harry Collins and Robert Evans, *Why Democracies Need Science* (Cambridge: Polity Press, 2017).

6. See, for example, Marcel LaFollette, *Making Science Our Own: Public Images of Science, 1910–1955* (Chicago: University of Chicago Press, 1990).

7. On evolution, see David L. Hull, *Darwin and His Critics: The Reception of Darwin's Theory of Evolution by the Scientific Community* (Cambridge, MA: Harvard University Press, 1973). On climate science, see Naomi Oreskes and Erik M. Conway, *Merchants of Doubt: How a Handful of Scientists Obscured the Truth on Issues from Tobacco Smoke to Global Warming* (New York: Bloomsbury, 2011). On the plural nature of scientific practice, see Peter Galison and David J. Stump, eds., *The Disunity of Science: Boundaries, Contexts, and Power* (Stanford, CA: Stanford University Press, 1996); Nancy Cartwright, *The Dappled World: A Study of the Boundaries of Science* (New York: Cambridge University Press, 1999); and Helen E. Longino, *The Fate of Knowledge* (Princeton, NJ: Princeton University Press, 2002), 175–202.

8. The status of science in American culture is well documented by Gieryn, *Cultural Boundaries of Science,* and Christopher P. Toumey, *Conjuring Science: Scientific Symbols and Cultural Meanings* (New Brunswick, NJ: Rutgers University Press, 1996).

9. See, for example, George E. DeBoer, *A History of Ideas in Science Education: Implications for Practice* (New York: Teachers College Press, 1991), and Scott L. Montgomery, *Minds for the Making: The Role of Science in American Education, 1750–1990* (New York: Guilford Press, 1994).

10. For education reform in general, see David Tyack and Larry Cuban, *Tinkering toward Utopia: A Century of Public School Reform* (Cambridge, MA: Harvard University Press, 1995).

11. In this way this story offers a contribution to work in the history of science and science studies that recognizes the circulation of ideas in science and culture in

contrast to linear models of knowledge diffusion (from science to the public), see, for example, Stephen Hilgartner, "The Dominant View of Popularization: Conceptual Problems, Political Uses," *Social Studies of Science* 20 (1990): 519–539; Roger Cooter and Stephen Pumfrey, "Separate Spheres and Public Places: Reflections on the History of Science Popularization and Science in Popular Culture," *History of Science* 32 (1994): 237–267; Greg Myers, "Discourse Studies of Scientific Popularization: Questioning the Boundaries," *Discourse Studies* 5 (2003): 265–279; and Lynn K. Nyhart, *Modern Nature: The Rise of the Biological Perspective in Germany* (Chicago: University of Chicago Press, 2009), 1–34.

1. FROM TEXTBOOK TO LABORATORY

1. *Official Directory of the World's Columbian Exposition, May 1st to October 30th, 1893: A Reference Book* (Chicago: W. B. Conkey, 1892), 37.

2. *Official Directory of the World's Columbian Exposition,* 37.

3. Richard Waterman Jr., "Educational Exhibits at the Columbian Exposition (I)," *Educational Review* 6 (1893): 274.

4. Waterman, "Educational Exhibits at the Columbian Exposition (I)," 268, 274.

5. Richard Waterman Jr., "Educational Exhibits at the Columbian Exposition (III)," *Educational Review* 7 (1894): 267; *Report of the Massachusetts Board of World's Fair Managers* (Boston: Wright & Potter, 1894), 104; Waterman, "Educational Exhibits at the Columbian Exposition (III)," 269–270. Rebecca B. Miller writes about the rise and fall of laboratory teaching with an eye toward questions of differentiation of science instruction in the United States in "Making Scientific Americans: Identifying and Educating Future Scientists and Nonscientists in the Early Twentieth Century," PhD diss., Harvard University, 2017.

6. Donald M. Scott, "The Popular Lecture and the Creation of a Public in Mid-Nineteenth-Century America," *Journal of American History* 66 (1980): 791–809; *Scientific American,* September 26, 1846, 1; Sally Gregory Kohlstedt, "Parlors, Primers, and Public Schooling: Education for Science in Nineteenth-Century America," *Isis* 81 (1990): 424–445; Robert V. Bruce, *The Launching of Modern American Science, 1846–1876* (Ithaca, NY: Cornell University Press, 1987), 254. Much of the early public interest in science is described in Bruce, *The Launching of Modern American Science,* 115–134, 339–356. Details about popularization during this period can be found in John C. Burnham, *How Superstition Won and Science Lost: Popularizing Science and Health in the United States* (New Brunswick, NJ: Rutgers University Press, 1987), 127–169.

7. "Professor Tyndall's First Lecture in America," *Scientific American,* November 2, 1872, 275; Bruce, *The Launching of Modern American Science,* 353–356. The importance of evolution to public interest in science is described by Daniel P. Thurs in

Science Talk: Changing Notions of Science in American Popular Culture (New Brunswick, NJ: Rutgers University Press, 2007), 53–89.

8. Note that science was taught in academies and private schools for a number of other reasons as well, including mental discipline and moral improvement, especially for girls, but this type of instruction was limited for the most part to the upper-class elite. See Kim Tolley, *The Science Education of American Girls: A Historical Perspective* (New York: RoutledgeFalmer, 2003). On the history of technology education in the United States, see Bruce, *The Launching of Modern American Science,* 150–165; and David F. Noble, *America by Design: Science, Technology, and the Rise of Corporate Capitalism* (New York: Oxford University Press, 1979). The public fascination with technology is described in Bruce, *The Launching of Modern American Science,* 127–134.

9. The reference to the work being an "educational classic" was made in Andrew W. Edson, "Things Most Worth While in Education," *Journal of Education* 86 (1917): 451; Herbert Spencer, "What Knowledge Is of Most Worth?," in *Education: Intellectual, Moral, and Physical* (London: G. Manwaring, 1861), 53.

10. Alexander Inglis, *Principles of Secondary Education* (Cambridge: Riverside Press, 1918), 512; "What Knowledge Is of Most Worth," *New York Times,* August 9, 1859, 3; "College Progress," *New York Times,* August 16, 1859, 4.

11. Legislative language from the Morrill Act as quoted in Edward Danforth Eddy Jr., *Colleges for Our Land and Time: The Land-Grant Idea in American Education* (New York: Harper and Brothers, 1957), 33. On the rise of science in higher education, see Jon H. Roberts and James Turner, *The Sacred and the Secular University* (Princeton, NJ: Princeton University Press, 2000); Bruce, *Launching of Modern American Science,* 326–338; and Andrew Jewett, *Science, Democracy, and the American University: From the Civil War to the Cold War* (New York: Cambridge University Press, 2012), 30–31. On science in the land grant institutions, see Roger Geiger, "The Rise and Fall of Useful Knowledge: Higher Education for Science, Agriculture & the Mechanic Arts, 1850–1875," *History of Higher Education Annual* 18 (1998): 47–65, and Eddy, *Colleges for Our Land and Time.*

12. The cultural proponents of science are the "scientific democrats" Andrew Jewett describes in *Science, Democracy, and the American University,* 21–81. Quotation from "Notices of New Books: The Culture Demanded by Modern Life," *New Englander and Yale Review* 27 (1868): 206; E. L. Youmans, ed., *The Culture Demanded by Modern Life: A Series of Addresses and Arguments on the Claims of Scientific Education* (New York: D. Appleton, 1867). Roger L. Geiger, *The History of American Higher Education: Learning and Culture from the Founding to World War II* (Princeton, NJ: Princeton University Press, 2014), 318–321; see also Robert E. Kohler, "The Ph.D. Machine: Building on the Collegiate Base," *Isis* 81 (1990): 645–649.

13. Geiger, "The Rise and Fall of Useful Knowledge," 59; Richard C. Maclaurin, "Science and Education," *School Review* 18 (1910): 319.

14. Simon Newcomb, "Abstract Science in America, 1776–1876," *North American Review* 122 (Jan. 1876): 122–123; Simon Newcomb, "What Is a Liberal Education?," *Science* n.s. 3 (1884): 435–436. For a detailed look at Newcomb's views on and public advocacy of scientific method, see Albert E. Moyer, *A Scientist's Voice in American Culture: Simon Newcomb and the Rhetoric of Scientific Method* (Berkeley: University of California Press, 1992), xi, 66–97.

15. "Teaching Physics," *Teacher* 25 (March 1872): 93; Frank P. Whitman, "The Beginnings of Laboratory Teaching in America" *Science* 8 (1898): 202. See also Geraldine Joncich, "Scientists and the Schools of the Nineteenth Century: The Case of American Physicists," *American Quarterly* 18 (1966): 667–685; S. A. Forbes, "History and Status of Public School Science Work in Illinois," in *Educational Papers by Illinois Science Teachers, 1889–1890* (Peoria, IL: J. W. Franks & Sons, 1891), 4.

16. William J. Reese, *The Origins of the American High School* (New Haven, CT: Yale University Press, 1995), 107–111. See also William J. Reese, "American High School Political Economy in the Nineteenth Century," *History of Education* 27 (1998): 255–265. There was always a population of students, particularly boys from more well-to-do families attending private academies, who studied Latin and Greek because mastering these classical languages both was required for college admission and, for those not planning on advanced study, was simply what it meant to be educated in the early 1800s. With few colleges open to women prior to 1850, however, female academies were free to offer more practical studies, and the sciences prominently filled this role for them, particularly subjects such as botany and chemistry. Over the course of the century, as more and more public high schools were established, the proportion of students preparing for college (which was small to begin with) declined further and further. On the differential course offerings for boys and girls during this period, see Tolley, *The Science Education of American Girls,* 35–53. On the common approaches to teaching at the time, see Reese, *The Origins of the American High School,* 139–140; Whitman, "The Beginnings of Laboratory Teaching in America," 201–206; Larry Owens, "Pure and Sound Government: Laboratories, Playing Fields, and Gymnasia in the Nineteenth-Century Search for Order," *Isis* 76 (1985): 185. See also William Torrey Harris, *How to Teach Natural Science in Public Schools,* 2nd ed. (Syracuse, NY: C. W. Bardeen, 1895). The use of objects as the starting point of instruction was common in the elementary grades in the Oswego movement and among Herbartian educators; see Harold B. Dunkel, *Herbart and Herbartianism: An Educational Ghost Story* (Chicago: University of Chicago Press, 1970), 243–244. And as late as 1889, the object method was still being advocated for students in the lower schools; see Samuel F. Clarke, William N. Rice, William G. Farlow, George MacLoskie, and C. O. Whitman, "Science in the Schools," *Education* 9 (1889): 547–549.

17. Harris, *How to Teach Natural Science in Public Schools,* 25; Chicago Board of Education, *Graded Course of Instruction for the Public Schools of Chicago,* 4th ed. (Chicago: Bryant, Walker & Craig, 1872), 23–24.

18. Conway Macmillan, "Current Methods in Botanical Instruction," *Education* 12 (1892): 463–464.

19. Jas. Lewis Howe, "The Teaching of Science," *Science* n.s. 19 (1892): 233.

20. William J. Reese, *Testing Wars in the Public Schools: A Forgotten History* (Cambridge, MA: Harvard University Press, 2013), 160–170; Forbes, "History and Status of Public School Science Work in Illinois," 7. On the pressure of teacher examinations in science, see Elmer Ellsworth Brown, *The Making of Our Middle Schools: An Account of the Development of Secondary Education in the United States* (New York: Longmans, Green, 1903), 416–417, and Heber Eliot Rumble, "*More* Science Instruction?," *Science Education* 33 (1949): 32–40. Teacher quoted in Forbes, "History and Status of Public School Science Work in Illinois," 14.

21. Reese, *Testing Wars in the Public Schools: A Forgotten History*, 205–208; Brown, *The Making of Our Middle Schools*, 416–417; J. Dorman Steele, *Fourteen Weeks in Zoology* (New York: A. S. Barnes, 1872), 5. For more on Steele and his textbooks, see Reese, *Origins of the American High School*, 105, 109, 122. See the various book advertisements published in John M. Hull, *Epitome of the New School Law of Illinois. In Force July 1st, 1872* (Chicago: Hadley Brothers, 1872).

22. Frank Wigglesworth Clarke, *A Report on the Teaching of Chemistry and Physics in the United States*, U.S. Bureau of Education Circular No. 6-1880 (Washington, DC: Government Printing Office, 1881), 6. Details of Clarke's report were provided in C. R. Mann, "Physics Teaching in the Secondary Schools of America," *Science* 30 (1909): 793.

23. Howe, "The Teaching of Science," 233; William Harmon Norton, "The Teaching of Science," *School Science* 2 (1902): 196.

24. "Information study" reference from Michigan Academy of Science, *A Booklet for Teachers of Zoology and Botany in the High Schools of Michigan* (Michigan: n.p., 1904), 5; Brown, *The Making of Our Middle Schools*, 416; Forbes, "History and Status of Public School Science Work in Illinois," 14 (emphasis in original).

25. Laurence R. Veysey, *The Emergence of the American University* (Chicago: University of Chicago Press, 1965), 125–127; Frederick Rudolph, *The American College and University: A History*, 2nd ed. (Athens: University of Georgia Press, 1990), 264–286; Bruce, *The Launching of Modern American Science*, 334–335; Owen Hannaway, "The German Model of Chemical Education in America: Ira Remsen at Johns Hopkins (1876–1913)," *Ambix* 23 (1976): 145–164. On the fame of Liebig's lab, see Ira Remsen, "Prof. Ira Remsen on Chemical Laboratories," *Nature* 49 (1894): 532; Whitman, "The Beginnings of Laboratory Teaching in America," 201–206; and Owens, "Pure and Sound Government," 183–185.

26. On the distinction between technical and liberal goals of lab work, see Josiah P. Cooke, *The Value and Limitations of Laboratory Practice in a Scheme of Liberal Education* (Cambridge, MA: Harvard University, 1892); Whitman, "The Beginnings of Laboratory Teaching in America," 203–204; Charles E. Bessey, "A

Summary History of the Department of Botany in the Iowa State College from 1870 to 1884," *The Alumnus* (October 1906): 11, box 1, folder 22, Charles E. Bessey Papers, RS 13/5/11, Special Collections Department, Iowa State University Library.

27. Frank P. Whitman wrote, "The striking features of the last twenty years have been the spread of science teaching by laboratory methods in the secondary schools, and the growth of university instruction in science, as distinguished from technical. The noble gift of Johns Hopkins, and its wise administration, began the latter movement, which resulted in the establishment of graduate schools all over the country." Whitman, "The Beginnings of Laboratory Teaching in America," 206; see also Roger L. Geiger, *The History of American Higher Education,* 340; Owens, "Pure and Sound Government," 184. On Remsen's background, see Hannaway, "The German Model of Chemical Education in America"; on Rowland, see Thomas C. Mendenhall, "Biographical Memoir of Henry Augustus Rowland, 1848–1901," in *Biographical Memoirs,* vol. V (Washington, DC: National Academy of Sciences, 1905), 115–140; Charles W. Eliot and Frank H. Storer, *A Manual of Inorganic Chemistry, Arranged to Facilitate the Experimental Demonstration of the Facts and Principles of the Science* (New York: Ivison, Blakeman, Taylor, 1876), v; Rufus P. Williams, "Teaching of Chemistry in Schools—1876, 1901," *Science* 14 (1901): 104 (emphasis in original). Charles Bessey made a similar statement in the preface to his *Botany for High Schools and Colleges* (New York: Henry Holt, 1880).

28. A. Hunter Dupree, "The Measuring Behavior of Americans," in *Nineteenth-Century American Science: A Reappraisal,* ed. George H. Daniels (Evanston, IL: Northwestern University Press, 1972), 22–37; Bruce, *The Launching of Modern American Science,* 64–74; Daniel J. Kevles, *The Physicists: The History of a Scientific Community in Modern America* (Cambridge, MA: Harvard University Press, 1987), 36–40; Daniel J. Kevles and Carolyn Harding, "The Physics, Mathematics, and Chemistry Communities in America, 1870–1915: A Statistical Survey," Social Science Working Paper no. 136, California Institute of Technology, 1977. See also John W. Servos, "Mathematics and the Physical Sciences in America, 1880–1930," *Isis* 77 (1986): 611–629. On the history of methodological writing in science, see Laurens Laudan, "Theories of Scientific Method from Plato to Mach: A Bibliographic Review," *History of Science* 7 (1968): 1–38. The widespread acceptance of the Baconian view of scientific method is described in, among other places, Burnham, *How Superstition Won and Science Lost,* 162–163; David L. Hull, "Charles Darwin and Nineteenth-Century Philosophies of Science," in *Foundations of Scientific Method: The Nineteenth Century,* ed. Ronald N. Giere and Richard S. Westfall (Bloomington: Indiana University Press, 1973), 115–117; and Richard R. Yeo, "Scientific Method and the Rhetoric of Science in Britain, 1830–1917," in *The Politics and Rhetoric of Scientific Method: Historical Studies,* ed. John A. Schuster and Richard R. Yeo (Dordrecht: D. Reidel, 1986), 263.

29. William T. Sedgwick, "Educational Value of the Methods of Science," *Educational Review* 5 (1893): 246. Descriptions of the inductive method were remarkably consistent from scientist to scientist; see, for example, William M. Davis, "Science in the Schools," *Educational Review* 13 (1897): 431. A detailed account of the American fascination with Bacon and how science in practice differed from strict Baconian prescriptions is found in S. A. Forbes, *The Method of Science and the Public School* (Champaign, IL: Gazette Press, 1901). On the changing commitment to induction among scientists, see Julie A. Reuben, *The Making of the Modern University: Intellectual Transformation and the Marginalization of Morality* (Chicago: University of Chicago Press, 1996), 36–50.

30. Remsen, "Prof. Ira Remsen on Chemical Laboratories," 533. Remsen reiterated these views years later in a talk before the Twentieth Century Club of Boston in 1902. Lyman C. Newell, "Professor Remsen on the Teaching of Science," *School Science* 2, no. 3 (1902): 129–132; William A. Locy, "On Teaching Zoology to College Classes," *Education* 9 (1889): 675; Angell quoted in Elliott R. Downing, "Report of Meetings: Upper Peninsula Educational Association," *School Science* 2 (1902): 368–369.

31. Charles W. Eliot, "What Is a Liberal Education?," *Century Magazine* 28 (June 1884): 209; Charles R. Barnes, "Sciences in the High School," *School Review* 6 (1898): 648; LaRoy F. Griffin, "The Laboratory in the School," *School and College* 1 (1892): 469.

32. Albert L. Arey, "The Educational Value of the Physical Sciences," paper presented at the meeting of the New York State Science Teachers' Association, Syracuse, NY, December 30, 1896, in Franklin W. Barrows, "The New York State Science Teachers' Association," *Science* 5 (1897): 461.

33. Charles W. Eliot, "The Laboratory Method of Teaching," *School Journal* 72 (1906): 213.

34. For references to mental discipline, see, for example, W. Whitman Bailey, "The Claims of Botany," *Education* 7 (1887): 706; Locy, "On Teaching Zoology to College Classes," 679; E. A. H. Allen, "The Teaching of Science: Science in Secondary Schools," *Education* 10 (1889): 108–117; and Charles F. Mabery, "Physical Science in the Secondary Schools," *Science* n.s. 21 (1893): 197. On the history of mental discipline, see Walter B. Kolesnik, *Mental Discipline in Modern Education* (Madison: University of Wisconsin Press, 1962).

35. Thomas R. Stowell, "The Educative Value of the Study of Biology," paper presented at the meeting of the New York State Science Teachers' Association, Syracuse, NY, December 31, 1896, in Franklin W. Barrows, "The New York States Science Teachers' Association," *Science* 5 (1897): 531–532; Griffin, "The Laboratory in the School," 469.

36. David Hollinger, "Inquiry and Uplift: Late Nineteenth-Century American Academics and the Moral Efficacy of Scientific Practice," in *The Authority of Experts:*

Studies in History and Theory, ed. Thomas L. Haskell (Bloomington: Indiana University Press, 1984), 143. See also Deborah J. Warner, "Physics as a Moral Discipline: Undergraduate Laboratories in the Late Nineteenth Century," *Rittenhouse* 6 (1992): 116–129; Hannaway, "The German Model of Chemical Education in America"; Jewett, *Science, Democracy, and the American University.* Michelle Hoffman traces this aspect of science education in the Canadian context in "Constructing School Science: Physics, Biology, and Chemistry Education in Ontario High Schools, 1880–1940," PhD diss., University of Toronto, 2013. See also Allan C. Story, "Report of the President," in Department of Public Instruction, City of Chicago, *Thirty-Third Annual Report of the Board of Education for the Year Ending, June 30, 1887* (Chicago: Jameson and Morse, 1888), 38; Eliot, "What Is a Liberal Education?," 209.

37. Charles W. Eliot, typescript beginning "The Report of the Committee of Ten has now been in the hands of the teachers of the country for about six months . . . ," 1894, box 241, Records of the President of Harvard University, Charles W. Eliot, 1869–1930, Harvard University Archives; William North Rice, "Science-Teaching in the Schools," *American Naturalist* 22 (1888): 770.

38. Henry A. Rowland, "The Physical Laboratory in Modern Education," in *The Physical Papers of Henry Augustus Rowland* (Baltimore, MD: The Johns Hopkins Press, 1902), 616; Norton, "The Teaching of Science," 196–197; Francis E. Lloyd and Maurice A. Bigelow, *The Teaching of Biology in the Secondary School* (New York: Longmans, Green, 1904), 10.

39. Rowland, "The Physical Laboratory in Modern Education," 617. For similar comments, see Sedgwick, "Educational Value of the Methods of Science," 253, and Griffin, "The Laboratory in the School," 477. On manhood in the nineteenth century, see E. Anthony Rotundo, *American Manhood: Transformations in Masculinity from the Revolution to the Modern Era* (New York: Basic Books, 1993), 167–193; see also Owens, "Pure and Sound Government," 182–194.

40. Sedgwick, "Educational Value of the Methods of Science," 245; Arey, "The Educational Value of the Physical Sciences," 461; J. M. Jameson, "Eastern Association of Physics Teachers," *Manual Training Magazine* 3 (July 1902): 233. On the manual-training movement in American schools, see Herbert M. Kliebard, *Schooled to Work: Vocationalism and the American Curriculum, 1876–1946* (New York: Teachers College Press, 1999), 1–25. T. J. Jackson Lears offers a nice discussion of the arts and crafts movement and educational reform in *No Place of Grace: Antimodernism and the Transformation of American Culture, 1880–1920* (Chicago: University of Chicago Press, 1994), 74–83. See also Owens, "Pure and Sound Government."

41. George W. Earle, "Address of the Vice-President," Report of the Thirty-Fourth Meeting of the Eastern Association of Physics Teachers, East Boston, MA, November 3, 1902, Eastern Association of Physics Teachers Records, 1896–1905, Niels Bohr Library, Center for the History of Physics, American Institute of Physics, College Park, MD (emphasis in original).

42. Eliot, "What Is a Liberal Education?," 209; Howe, "Teaching of Science," 235. See also Barnes, "Sciences in the High School," 647; Clarence M. Weed, "On the Use of the Laboratory Note Book," *Education* 16 (1896): 546; and Louis Murbach, "Method in Science Teaching," *School Science* 2 (1902): 12. Charles Bessey had made this the central theme of his address before the newly formed NEA Department of Science Instruction in 1896, imploring those assembled, "Let us hear less in the schools of the practical value of science. Let us emphasize its vastly greater importance in the making of man." Charles E. Bessey, "Presidential Address—Science and Culture," in National Educational Association, *Journal of Proceedings and Addresses of the Thirty-Fifth Annual Meeting* (Chicago: University of Chicago Press, 1896), 942.

43. Louis Menand, *The Metaphysical Club: A Story of Ideas in America* (New York: Farrar, Straus and Giroux, 2001); Owens, "Pure and Sound Government"; Hollinger, "Inquiry and Uplift"; Carl F. Kaestle, *Pillars of the Republic: Common Schools and American Society, 1780–1860* (New York: Hill and Wang, 1983).

2. THE LABORATORY IN PRACTICE

1. "Kansas City's University: The Splendid New High School Building Nearing Completion," *Kansas City Times,* November 19, 1893, 11.

2. "A Model High School: Fame of the One for Brooklyn Girls Widespread," *New York Times,* April 7, 1895, 30; George H. Gerkheide, "New Biology Laboratory: That at the Lake View High School Will Be a Fine One," *Chicago Daily Tribune,* January 17, 1895, 5; and "High School Architect Discusses Plans," *San Jose Mercury,* February 7, 1907, 9. Similar comments were made in "The New High School Building at Prairie du Chien," *Wisconsin Journal of Education* 27 (1897): 55; "Santa Barbara: Times Man Visits the High School," *Los Angeles Times,* November 12, 1891; and "Education: Schools That Are Unexcelled in Any Part of America," *Los Angeles Times,* December 4, 1891. On the civic importance of high schools in the nineteenth century, see William J. Reese, *The Origins of the American High School* (New Haven, CT: Yale University Press, 1995), 80–102; and William W. Cutler, III, "Cathedral of Culture: The Schoolhouse in American Educational Thought and Practice since 1820," *History of Education Quarterly* 29 (1989): 1–40.

3. "A Model High School," 30.

4. Robert V. Bruce, *The Launching of Modern American Science, 1846–1876* (Ithaca, NY: Cornell University Press, 1987), 335–336; Henry James, *Charles W. Eliot: President of Harvard University, 1869–1909* (Boston: Houghton Mifflin, 1930), 1:160–165; Tenney L. Davis, "Eliot and Storer: Pioneers in the Laboratory Teaching of Chemistry," *Journal of Chemical Education* 6 (1929): 868–879.

5. Charles W. Eliot to Edwin H. Hall, March 2, 1881, and April 14, 1881, Item 300, Edwin Herbert Hall Papers, Houghton Library, Harvard University. The Harvard physics department at the time was staffed by Joseph Lovering, who was an ac-

complished lecturer but had no bent for research of any kind, and John Trowbridge, who was committed to physics research and was the first to provide students with opportunities for original physical research, though not routine elementary laboratory instruction of the sort found in chemistry. See Edwin H. Hall, "Biographical Memoir of John Trowbridge, 1843–1923," *Biographical Memoirs,* vol. XIV (Washington, DC: National Academy of Sciences, 1931).

6. Edwin H. Hall, "Teaching Elementary Physics, II," *Educational Review* 5 (1893): 325; Harvard University, *Harvard University Catalogue, 1886–87* (Cambridge, MA: Harvard University, 1886), 77; Edwin H. Hall, "A Story of Experience with Physics as a Requirement for Admission to College, with Certain Propositions," *Proceedings of the Third Annual Conference of the New York State Science Teachers' Association, 1898* (Albany: University of the State of New York, 1899), 584–589; Albert E. Moyer, "Edwin Hall and the Emergence of the Laboratory in Teaching Physics," *Physics Teacher* 14 (1976): 96–103. On the change in overall admissions requirements, see Charles W. Eliot, "President's Report for 1885–86," in *Annual Reports of the President and Treasurer of Harvard College, 1885–86* (Cambridge, MA: Harvard University, 1887), 7–8, and "Modern Ideas at Harvard: The System of Examinations for Admission Revised," *New York Times,* February 19, 1885. On the dissatisfaction with previous examinations, see Charles W. Eliot, "President's Report for 1886–87," in *Annual Reports of the President and Treasurer of Harvard College, 1886–87* (Cambridge, MA: Harvard University, 1888), 9, and Edwin H. Hall, "Physics Teaching at Harvard Fifty Years Ago," *American Physics Teacher* 6 (1938): 17–20.

7. John Trowbridge, *The New Physics: A Manual of Experimental Study for High Schools and Preparatory Schools for College* (New York: D. Appleton, 1884); Edward C. Pickering, *Elements of Physical Manipulation* (Boston: Houghton, Mifflin, 1882); A. M. Worthington, *A First Course of Physical Laboratory Practice* (London: Rivingtons, 1886); Alfred P. Gage, *A Text-Book on the Elements of Physics for High Schools and Academies* (Boston: Ginn, Heath, 1883).

8. Edwin H. Hall and Joseph Y. Bergen, *A Text-Book of Physics, Largely Experimental: On the Basis of the Harvard College "Descriptive List of Elementary Physical Experiments"* (New York: Henry Holt, 1891), iii; Joseph Lovering, "The Jefferson Physical Laboratory," in *Annual Reports of the President and Treasurer of Harvard College, 1886–87* (Cambridge, MA: Harvard University, 1888), 138. Textbook list given in Hall to Webster, September 9, 1937, box 104, Records of the Harvard University Department of Physics, Harvard University Archives. Harvard University, *Descriptive List of Elementary Physical Experiments. Intended for Use in Preparing Students for Harvard College* (Cambridge, MA: Harvard University, 1889).

9. Edwin H. Hall, "Admission Physics—Elementary, Introduction," manuscript copy of the Harvard Descriptive List, 1887, box 104, Records of the Harvard University Department of Physics, Harvard University Archives. Details of the

exercises can be found in Harvard University, *Descriptive List of Elementary Physical Experiments*. Edwin H. Hall, "Experimental Physics for Schools," *Science* n.s. 10 (1887): 129–130.

10. Harvard University, *Descriptive List of Elementary Physical Experiments*, 7. The importance of laboratory notebooks was a topic for numerous articles in education journals of the time. See, for example, LaRoy Griffin, "The Laboratory in the School," *School and College* 1 (1892): 473, and Franklin Turner Jones, "A Method of Keeping Laboratory Notes," *School Science* 3 (1904): 449–450.

11. Hall, "Experimental Physics for Schools," 129; Hall and Bergen, *A Text-Book of Physics*. The goals of laboratory instruction are from the introduction to the 1889 pamphlet of the Descriptive List, which is reproduced in full in this textbook, vii; Edwin H. Hall and Joseph Y. Bergen, *A Text-Book of Physics, Largely Experimental: Including the Harvard College "Descriptive List of Elementary Exercises in Physics*, rev. ed. (New York: Henry Holt, 1897), 3; Edwin H. Hall, "Teaching Elementary Physics," *Educational Review* 4 (1892): 160.

12. See Graeme J. N. Gooday, *The Morals of Measurement: Accuracy, Irony, and Trust in Late Victorian Electrical Practice* (New York: Cambridge University Press, 2004), and Iwan Rhys Morus, *When Physics Became King* (Chicago: University of Chicago Press, 2005), 226–260. Gooday has argued convincingly that the rush of British industrial development in the second half of the nineteenth century was largely responsible for the rise of measurement-based laboratory physics in the United Kingdom and its use in training teachers and engineers of the period; see Graeme Gooday, "Precision Measurement and the Genesis of Physics Teaching Laboratories in Victorian Britain," *British Journal for the History of Science* 23 (1990): 25–51. Precision measurement was a hallmark of the German science seminars for teachers as well; Kathryn M. Olesko, "German Models, American Ways: The 'New Movement' among American Physics Teachers, 1905–1909," in *German Influences on Education in the United States to 1917*, ed. Henry Geitz, Jürgen Heideking, and Jurgen Herbst (Cambridge: Cambridge University Press, 1995), 129–153. Michelson's remark was made in a convocation address at the University of Chicago in 1894; Albert A. Michelson, "Some of the Objects and Methods of Physical Science," *Quarterly Calendar* 3, no. 2 (August 1894): 15.

13. Thomas C. Mendenhall, "Biographical Memoir of Henry Augustus Rowland, 1848–1901," *Biographical Memoirs*, vol. V (Washington, DC: National Academy of Sciences, 1905), 126. On the overall character of American physical science during this period, see John W. Servos, "Mathematics and the Physical Sciences in America, 1880–1930," in *The Scientific Enterprise in America: Readings from Isis*, ed. Ronald L. Numbers and Charles E. Rosenberg (Chicago: University of Chicago Press, 1996), 141–144, and Albert E. Moyer, "American Physics in 1887," in *The Michelson Era in American Science: 1870–1930*, ed. Stanley Goldberg and Roger H. Stuewer (New York: American Institute of Physics, 1988), 104–106. P. W. Bridgman, "Biograph-

ical Memoir of Edwin Herbert Hall, 1855–1938," in National Academy of Sciences, *Biographical Memoirs,* vol. XXI (1939), 82–84.

14. On Gage's work, see Steven C. Turner, "Changing Images of the Inclined Plane: A Case Study of a Revolution in American Science Education," *Science and Education* 21 (2012): 245–270. On the objections expressed to laboratory work in physics due to the cost of equipment, see the author's preface to Gage, *Text-Book on the Elements of Physics;* also Rufus P. Williams, "Alfred Payson Gage," *School Science* 3 (1903): 49–52. Edwin H. Hall, "Biographical Memoir of Wallace C. Sabine," typescript, item 161, Edwin Hall Papers, Houghton Library, Harvard University. See also the statement by John Trowbridge, the director of the Jefferson Physical Laboratory, on the new low-cost apparatus designed by Hall and Sabine: John Trowbridge, "The Jefferson Physical Laboratory," in *Annual Reports of the President and Treasurer of Harvard College, 1890–91* (Cambridge, MA: Harvard University, 1892), 169–171.

15. Quotation from E. D. Cope and J. S. Kingsley, "Editorials," *American Naturalist* 26 (1893): 1014–1015. John Trowbridge, "The Jefferson Physical Laboratory," *Annual Reports of the President and Treasurer of Harvard College, 1888–89* (Cambridge, MA: The University, 1890), 168; Charles W. Eliot, "President's Report for 1888–89," *Annual Reports of the President and Treasurer of Harvard College, 1889–90* (Cambridge, MA: Harvard University, 1891), 31.

16. James, *Charles W. Eliot,* 365, 351, 365; Hall, "Experimental Physics for Schools," 130.

17. Agassiz's famous Anderson School of Natural History promoted the study of nature directly, but it ultimately led to a curriculum movement in nature study that not only was focused on elementary schools but also included a wide range of goals for students beyond understanding the processes of science—everything from rural life to aesthetic appreciation; see Kim Tolley, *The Science Education of American Girls: A Historical Perspective* (New York: RoutledgeFalmer, 2003), 130, 135–136. Sally Gregory Kohlstedt, "Nature, Not Books: Scientists and the Origins of the Nature-Study Movement in the 1890s," *Isis* 96 (2005): 324–352. On the inspirational force of Agassiz, see John Pendleton Campbell, *Biological Teaching in the Colleges of the United States,* Bureau of Education Circular of Information No. 9 (Washington, DC: Government Printing Office, 1891), 126–128.

18. Charles E. Bessey, ed., "Botany," *American Naturalist* 15 (1881): 732; see also Raymond J. Pool, "A Brief Sketch of the Life and Work of Charles Edwin Bessey," *American Journal of Botany* 2 (1915): 505–518.

19. Charles E. Bessey, *Botany for High Schools and Colleges* (New York: Henry Holt, 1880); Charles E. Bessey, *The Essentials of Botany* (New York: Henry Holt, 1884); John M. Coulter, review of *Botany for High Schools and Colleges,* by Charles E. Bessey, *Botanical Gazette* 5 (1880): 96; Conway Macmillan, "Current Methods in Botanical Instruction," *Education* 12 (1892): 467. Macmillan notes the German

influence on Bessey. On Bessey's professional background and involvement in educa-tion reform, see ch. 6 of Thomas R. Walsh, "Charles E. Bessey: Land-Grant College Professor," PhD diss., University of Nebraska–Lincoln, 1972; Richard A. Overfield, "Charles E. Bessey: The Impact of the 'New' Botany on American Agriculture, 1880–1910," *Technology and Culture* 16 (1975): 162–181; and Ronald C. Tobey, *Saving the Prairies: The Life Cycle of the Founding School of American Plant Ecology, 1895–1955* (Berkeley: University of California Press, 1981), 24–47.

20. On efforts to communicate science to the public in Britain, see Bernard Lightman, *Victorian Popularizers of Science: Designing Nature for New Audiences* (Chicago: University of Chicago Press, 2007). T. H. Huxley and H. N. Martin, *A Course of Practical Instruction in Elementary Biology* (London: Macmillan, 1876). Subsequent books that emulated Huxley's plan include A. S. Packard, *Zoology for High Schools and Colleges* (New York: Henry Holt, 1879); Buel P. Colton, *Elementary Course in Practical Zoölogy* (Boston: D. C. Heath, 1886); Emanuel R. Boyer, *A Labora-tory Manual in Elementary Biology: An Inductive Study in Animal and Plant Morphology: Designed for Preparatory and High Schools* (Boston: D. C. Heath, 1894); and Charles Wright Dodge, *Introduction to Elementary Practical Biology: A Laboratory Guide for High-School and College Students* (New York: Harper and Brothers, 1894). On the in-fluence of Huxley on American biology instruction, see J. T. Rothrock, "Home and Foreign Modes of Teaching Botany, III," *Botanical Gazette* 6 (1881): 235; Campbell, *Biological Teaching in the Colleges of the United States,* 129; and Francis E. Lloyd and Maurice A. Bigelow, *The Teaching of Biology in the Secondary School* (New York: Long-mans, Green, 1904), 295, 352–353. The influence of both Huxley and Bessey is de-tailed in John M. Coulter, "Laboratory Courses of Instruction," *Botanical Gazette* 10 (1885): 417–421. The focus of most laboratory work in biology in the later 1880s and 1890s was on the analysis of ideal-type organisms through dissection. See Lloyd and Bigelow, *The Teaching of Biology in the Secondary School;* George W. Hunter, *Science Teaching at Junior and Senior High School Levels* (New York: American Book Com-pany, 1934), 27–28; Henry R. Linville, "Old and New Ideals in Biology Teaching," *School Science and Mathematics* 10 (1910): 210–216.

21. W. Whitman Bailey, "The Claims of Botany," *Education* 7 (1887): 705; William A. Locy, "The Teaching of Science: On Teaching Zoölogy to College Classes," *Edu-cation* 9 (1889): 673.

22. Quotation from Edwin H. Hall, *Elementary Lessons in Physics: Mechanics (In-cluding Hydrostatics) and Light* (New York: Henry Holt, 1894), iv–v. On the unique characteristics of natural history and the shifting research priorities in academic bi-ology at the end of the nineteenth century, see Keith R. Benson, "From Museum Research to Laboratory Research: The Transformation of Natural History into Aca-demic Biology," in *The American Development of Biology,* ed. R. Rainger, K. R. Benson, and J. Maienschein (Philadelphia: University of Pennsylvania Press, 1988), 49–83, and Lynn K. Nyhart, "Natural History and the 'New' Biology," in *Cultures of Natural*

History, ed. N. Jardine, J. A. Secord, and E. C. Spary (Cambridge: Cambridge University Press, 1996), 426–443. The idea of visual epistemology is an adaptation of Steven Conn's idea of "object-based" epistemology; see his *Museums and American Intellectual Life, 1876–1926* (Chicago: University of Chicago Press, 2000), 4–9.

23. Bessey, *Botany for High Schools and Colleges,* iii; Bessey, *The Essentials of Botany,* iv. The Nebraska High School Manual, which Bessey helped write, insisted that "there should be in every high school six microscopes, each with its accompanying accessories." Charles E. Bessey, "High School Botany," *Science* 7 (1898): 266.

24. Linville, "Old and New Ideals in Biology Teaching," 210; Dodge, *Introduction to Elementary Practical Biology,* 7–121. On the role of evolutionary theory as an organizing framework for biology textbooks in the early twentieth century, see Philip J. Pauly, "The Development of High School Biology: New York City, 1900–1925," *Isis* 82 (1991): 662–688.

25. On the parenthetical point, see Charles H. Ham, *Manual Training: The Solution of Social and Industrial Problems* (New York: Harper and Brothers, 1886), ch. 4. Quotation from Locy, "The Teaching of Science," 675–676. Dodge, *Introduction to Elementary Practical Biology,* xx–xxi. The importance of the laboratory notebook for keeping drawings and descriptions in biology is discussed in Clarence M. Weed, "On the Use of the Laboratory Note Book," *Education* 16 (1896): 546–549. A similar restriction on the number of illustrations was found in A. Milnes Marshall and C. Herbert Hurst, *A Junior Course of Practical Zoology,* 5th ed., revised by F. W. Gamble (New York: G. P. Putnam's Sons, 1899).

26. MacMillan, "Current Methods in Botanical Instruction," 468. On the professionalization of botany during this period, see Elizabeth B. Keeney, *The Botanizers: Amateur Scientists in Nineteenth-Century America* (Chapel Hill: University of North Carolina Press, 1992), and Richard A. Overfield, *Science with Practice: Charles E. Bessey and the Maturing of American Botany* (Ames: Iowa State University Press, 1993).

27. Details of the publicity and controversies surrounding the Committee of Ten and its report can be found in Edward A. Krug, *The Shaping of the American High School, 1880–1920* (New York: Harper & Row, 1964), 39–65; see also Marc A. VanOverbeke, *The Standardization of American Schooling: Linking Secondary and Higher Education, 1870–1910* (New York: Palgrave Macmillan, 2008).

28. National Education Association, *Report of the Committee of Ten on Secondary School Studies with the Reports of the Conferences Arranged by Committee* (New York: American Book Company, 1894), 5.

29. National Education Association, *Report of the Committee of Ten,* 119; Hall to Eliot, May 4, 1893, box 43, Records of the President of Harvard University, Charles W. Eliot, 1869–1930, Harvard University Archives.

30. National Education Association, *Report of the Committee of Ten,* 139, 211.

31. National Education Association, *Report of the Committee of Ten,* 18.

32. "The Report on Secondary Education," *Dial* 16 (1894): 37. See also "What Children Should Study: A Learned Committee Suggests Many Educational Changes," *New York Times,* January 1, 1894, and "A Big Question for Teachers: Important Report before Asbury Park Convention," *New York Times,* July 13, 1894. The comment on the influence of the report is from Elmer Ellsworth Brown, *The Making of Our Middle Schools: An Account of the Development of Secondary Education in the United States* (New York: Longmans, Green, 1903), 384. G. C. Bush, "The Status of the Physical Sciences in the High School," *School Science and Mathematics* 5 (1905): 432; "A Big Question for Teachers," *New York Times;* Herbert G. Espy, *The Public Secondary School: A Critical Analysis of Secondary Education in the United States* (Boston: Houghton Mifflin, 1939), 240.

33. A. F. Nightingale et al., "Report of the Committee on College Entrance Requirements," in National Education Association, *Journal of Proceedings and Addresses of the Thirty-Eighth Annual Meeting* (Chicago: University of Chicago Press, 1899), 651–655; E. H. Hall, "Outline of Laboratory Work in Physics for Secondary Schools," in National Education Association, *Journal of Proceedings and Addresses of the Thirty-Eighth Annual Meeting* (Chicago: University of Chicago Press, 1899), 809–811.

34. Hall, "Teaching Elementary Physics, II," 333; W. J. Chase and C. H. Thurber, "Tabular Statement of Entrance Requirements to Representative Colleges and Universities of the United States," *School Review* 4 (1896): 341–412; Charles S. Palmer, "Résumé and Critique of the Tabulated College Requirements in Natural Sciences," *School Review* 4 (1896): 452–460.

35. Krug, *The Shaping of the American High School,* 146–151. See also Brown, *The Making of Our Middle Schools,* 388–389. Eliot's long support for the board is noted in James, *Charles W. Eliot,* 368. College Entrance Examination Board, *Examination Questions in Botany, Drawing, Chemistry, Geography, Physics, 1901–1905* (Boston: Ginn, 1905). The number of laboratory notebooks graded is given in College Entrance Examination Board, *Second Annual Report of the Secretary, 1902* (New York: The Board, 1902), 28; College Entrance Examination Board, *Sixth Annual Report of the Secretary* (New York: The Board, 1906), 36.

36. Edwin H. Hall, "The Relations of Colleges to Secondary Schools in Respect to Physics," *Science* 30 (1909): 581; C. R. Mann, "Physics Teaching in the Secondary Schools of America," *Science* 30 (1909): 793.

37. Data on this is taken from the City of Chicago *Annual Reports of the Board of Education.* On the shift to laboratories in the high schools during this period, see Dale Allen Gyure, "The Transformation of the Schoolhouse: American Secondary School Architecture and Educational Reform, 1880–1920," PhD diss., University of Virginia, 2001, 61.

38. D. R. Cameron, Geo. L. Warner, and Geo. W. Stanford, "Report of Committee on High Schools," Public Schools of the City of Chicago, *Thirty-Ninth Annual Report of the Board of Education for the Year Ending June 30, 1893* (Chicago:

Geo. K. Hazlit, 1894), 116; Albert G. Lane, "Report of the Superintendent," Public Schools of the City of Chicago, *Fortieth Annual Report of the Board of Education for the Year Ending June 29, 1894* (Chicago: J. M. W. Jones, 1895), 45; A. F. Nightingale, "Supplementary Reports: High Schools," Public Schools of the City of Chicago, *Forty-First Annual Report of the Board of Education for the Year Ending June 28, 1895* (Chicago: J. M. W. Jones, 1895), 93.

39. Albert G. Lane, "Report of the Superintendent of Schools," Public Schools of the City of Chicago, *Forty-Second Annual Report of the Board of Education for the Year Ending June 26, 1896* (Chicago: J. M. W. Jones, 1896), 73; report of the meeting of the Central Association of Physics Teachers, meeting on November 28, 1902, comments by Ernest J. Andrews, published in "The Central Association of Physics Teachers," *School Science* 2 (1903): 478–479.

40. David L. Webster, "Contributions of Edwin Herbert Hall to the Teaching of Physics," *American Physics Teacher* 6 (1938): 15; see also "School Equipment: Improved Scientific Instruments," *School Journal* 50 (1895): 275, and Trowbridge, "The Jefferson Physical Library," *Annual Reports of the President and Treasurer of Harvard College, 1890–91,* 170. Steven Turner questions whether the Harvard physics course was, in fact, the one that was widely adopted in the United States, arguing that non-Harvard apparatus and lists of experiments were also widely used; see Turner, "Changing Images of the Inclined Plane." The point I make here is only that a laboratory-based course with a quantitative emphasis did indeed become the norm and that many science educators at the time credited Hall with that development, not that Hall's course was the exact model that was widely implemented in schools. Edwin H. Hall, "Grammar-School Physics," *Educational Review* 6 (1893): 242–248. The course of experiments is described in City of Cambridge, *Annual Report of the School Committee, 1894* (Boston: Cashman, O'Connor, 1895), 101–102. On the general advocacy of laboratory methods in the lower grades, see Charles B. Scott, "Laboratory Methods in Elementary Schools," *Journal of Proceedings and Addresses: Session of the Year 1894* (St. Paul, MN: National Education Association, 1895), 191–197.

41. National Education Association, *Report of the Committee of Ten,* 169; Adelia R. Hornbrook, *Laboratory Methods of Teaching Mathematics in Secondary Schools* (New York: American Book Company, 1895); G. W. Myers, "The Laboratory Method in the Secondary School," *School Review* 11 (1903): 727–741; and W. Betz, "The Laboratory Method of Teaching Mathematics," in *Proceedings of the Ninth Annual Conference of the New York State Science Teachers' Association, 1904* (Albany: New York State Education Department, 1905), 118–121. The equipment list is found in Myers. Plummer's comments are in "Modern Education: Its Principal Phases from Kindergarten to University Discussed by the Women's Club," *Brooklyn Eagle,* October 27, 1896. A. B. Prescott, "The American Association, President's Address," *Nature* 46 (1892): 411. See also Mary E. Wilder, *The Study of History by the Laboratory Method for High and Grammar Schools: England* (Boston: Lee and Shepard, 1895).

42. Frank P. Whitman, "The Beginnings of Laboratory Teaching in America," *Science* 8 (1898): 201–206.

3. STUDENT INTEREST AND THE NEW MOVEMENT

1. Dorothy Ross, *G. Stanley Hall: The Psychologist as Prophet* (Chicago: University of Chicago Press, 1972).

2. Richard Waterman Jr., "International Educational Congresses of 1893," *Educational Review* 6 (1893): 160–161. The overall description of the congress is given in this review. For a detailed outline of the session, see "Department Congress of Experimental Psychology: Secretary's Report," *Proceedings of the International Congress of Education of the World's Columbian Exposition, Chicago, July 25–28, 1893, under the Charge of the National Education Association of the United States* (New York: National Education Association, 1894), 713–716.

3. See Carl F. Kaestle, *Pillars of the Republic: Common Schools and American Society, 1780–1860* (New York: Hill and Wang, 1983).

4. J. C. Packard, A. B. Kimball, M. S. Powers, H. J. Chase, C. F. Warner, C. R. Allen, and Frank Rollins, open letter to Boston-area physics teachers, January 1, 1895 (emphasis in original), Eastern Association of Physics Teachers Records, 1896–1905, Niels Bohr Library, American Institute of Physics, College Park, MD (hereafter EAPT Records); Constitution of the Eastern Association of Physics Teachers (adopted March 2, 1895), EAPT Records.

5. Hall's talk is summarized in "Report of the Fifteenth Meeting of the Eastern Association of Physics Teachers," November 6, 1897, EAPT Records. *Plans and Specifications of Physical Laboratories (Ideal and Actual) for Secondary Schools, A Committee Report to the Eastern Association of Physics Teachers* (Cambridgeport, MA: Eastern Association of Physics Teachers, 1898), 3; EAPT, *Statistics Concerning Instruction in Physics in Eastern Secondary Schools* (Bridgewater, MA: EAPT, 1899), 4, EAPT Records.

6. Charles W. Warner, "An Effective Order of Topics in Teaching Physics," paper read before the Eastern Association of Physics Teachers, February 8, 1896, EAPT Records; "Report of the Eighteenth Meeting of the Eastern Association of Physics Teachers," March 26, 1898 (emphasis in original), EAPT Records.

7. J. C. Packard, "The Place of the Lecture in Teaching Elementary Physics," Report of the Twenty-Fifth Meeting of the Eastern Association of Physics Teachers, October 28, 1899, EAPT Records; W. D. Jackson, "The Quiz or Recitation Method of Teaching Physics," Report of the Twenty-Fifth Meeting of the Eastern Association of Physics Teachers, October 28, 1899, EAPT Records; Eastern Association of Physics Teachers, *Methods of Instruction on Physics in Secondary Schools: A Committee Report to the Eastern Association of Physics Teachers* (Boston: Eastern Association of Physics Teachers, 1900), 9, 13.

8. Address of G. Stanley Hall, "Correlation of Manual Training and Physics," Report of the Thirty-Third Meeting of the Eastern Association of Physics Teachers, May 24, 1902, p. 4, EAPT Records. Later published as G. Stanley Hall, "Some Criticism of High-School Physics, and Manual-Training and Mechanic-Arts High Schools, with Suggested Correlations," *Manual Training Magazine* 3 (1902): 189–200.

9. Hall, "Correlation of Manual Training and Physics," 4–5.

10. G. Stanley Hall, "How Far Is the Present High-School and Early College Training Adapted to the Nature and Needs of Adolescents?," *School Review* 9 (1901): 651–652. Hall's talk and the various rebuttals—by Charles Eliot in particular, who was in attendance—were reported in the *New York Times;* see "Education of the Boy: High School and Early College Training Discussed," *New York Times,* October 13, 1901.

11. John M. O'Donnell, *The Origins of Behaviorism: American Psychology, 1870– 1920* (New York: New York University, 1985), 1, 26–31, 91–105; "The American Psychological Association," *Science* n.s. 21 (1893): 34.

12. On this background, see Ellen Condliffe Lagemann, *An Elusive Science: The Troubling History of Education Research* (Chicago: University of Chicago Press, 2000), 23–32. See also Alice Boardman Smuts, *Science in the Service of Children, 1893–1935* (New Haven, CT: Yale University Press, 2006), 31–48.

13. Hugo Münsterberg, "The Danger from Experimental Psychology," *Atlantic Monthly* 81 (1898): 159. Similar sentiments were expressed in Hugo Münsterberg, "Psychology and the Real Life," *Atlantic Monthly* 81 (1898): 602–613, and Josiah Royce, "The New Psychology and the Consulting Psychologist," *The Forum* 26 (1898): 80–96. Similar reservations were expressed by William James in *Talks to Teachers on Psychology and to Students on Some of Life's Ideals* (New York: Henry Holt, 1899), 7–8. On the prominence of child study in the new psychology and its connection to evolution, see G. Stanley Hall, "The New Psychology," *Harper's Monthly* 103 (1901): 731–732. The practical applications of psychology figured significantly in its development as a profession; O'Donnell, *Origins of Behaviorism,* 118–122. See also Lagemann, *An Elusive Science,* 23–39.

14. Hall, "The New Psychology," 732; Ross, *G. Stanley Hall,* 279–308.

15. Enrollment data from *Reports of the Commissioner of Education.* United States population data from U.S. Census records. The report on the "High School Movement" was written by Elmer Ellsworth Brown. See Elmer Ellsworth Brown, "Secondary Education," *Report of the Commissioner of Education for the Year 1903* (Washington, DC: Government Printing Office, 1905), 1:563–583. On the pace of high school building, see Alexander Inglis, "Secondary Education," in *Twenty-Five Years of American Education: Collected Essays,* ed. I. L. Kandel (New York: Macmillan, 1924), 251–253. The figure of one high school per day is for the period from 1890 to 1918.

16. William J. Fisher, "The Drift in Secondary Education," *Science* 36 (1912): 590; Charles H. Judd, "Meaning of Science in Secondary Schools," *School Science and Mathematics* 12 (1912): 88. On the demographics of the student body, see Fred D. Barber, *The Physical Sciences in Our Public Schools* (Normal: Illinois State Normal University, 1913), 13. Contrary to conventional wisdom, high schools had long emphasized practical over academic studies; Edward A. Krug, *The Shaping of the American High School, 1880–1920* (New York: Harper & Row, 1964), 7. On the long-standing practical orientation of high schools, see William J. Reese, *The Origins of the American High School* (New Haven, CT: Yale University Press, 1995), 260, and William J. Reese, "American High School Political Economy in the Nineteenth Century," *History of Education* 27, no. 3 (1998): 255–265.

17. G. Stanley Hall, *Adolescence: Its Psychology and Its Relations to Physiology, Anthropology, Sociology, Sex, Crime, Religion, and Education* (New York: D. Appleton, 1904), II:510, 156, 154–159.

18. C. R. Mann, C. H. Smith, and C. F. Adams, "A New Movement among Physics Teachers," *School Review* 14 (1906): 212. On the establishment of CASMT, see Central Association of Science and Mathematics Teachers, *A Half Century of Science and Mathematics Teaching* (Oak Park, IL: Central Association of Science and Mathematics Teachers, 1950).

19. Frank M. Gilley, Ernest R. von Nardoff, and W. E. Tower, "Report of Department Committee on Physics Courses: First-Year Course in Physics Adopted by the Department of Science Instruction of the National Education Association, 1905," *Journal of Proceedings and Addresses of the Forty-Fourth Annual Meeting, 1905* (Winona, MN: National Educational Association, 1905), 815–817. A summary of the movement with details from each circular is given in Gilbert Random, "The So-Called New Movement among Physics Teachers," *Western Journal of Education* 13 (October 1908): 551–559. See also Kathryn M. Olesko, "German Models, American Ways: The 'New Movement' among American Physics Teachers, 1905–1909," in *German Influences on Education in the United States to 1917,* ed. Henry Geitz, Jürgen Heideking, and Jurgen Herbst (New York: Cambridge University Press, 1995), 129–153.

20. C. R. Mann, "The New Movement among Physics Teachers: Circular IV," *School Science and Mathematics* 6 (1906): 787–789. Quotation is from C. R. Mann, "A New Movement among Physics Teachers: Circular III," *School Science and Mathematics* 6 (1906): 700.

21. C. R. Mann, "On Science Teaching: V," *School Science and Mathematics* 6 (1906): 197; H. L. Terry, "The New Movement in Physics Teaching," *Educational Review* 37 (1909): 13; C. R. Mann, "The Aims and Tendencies in Physics Teaching," *School Science and Mathematics* 6 (1906): 724.

22. C. R. Mann, "The Meaning of the Movement for the Reform of Science Teaching," *Educational Review* 34 (1907): 14; C. Riborg Mann, *The Teaching of Physics for Purposes of General Education* (New York: Macmillan, 1912), 178. Mann's

repeated references to psychology can be found in C. R. Mann, "The New Movement for the Reform of Physics Teaching in Germany, France, and America," in *Proceedings of the Eleventh Annual Conference of the New York State Science Teachers' Association, 1906* (Albany: New York State Education Department, 1907), 94; and Mann, "On Science Teaching: V," 196–197.

23. C. R. Mann, "The New Movement among Physics Teachers—Circular VI," *School Science and Mathematics* 8 (1908): 522.

24. Nicholas Murray Butler, "Physics Teaching in Secondary Schools," *Educational Review* 37 (1909): 88; John F. Woodhull, "Symposium on the Purpose and Organization of Physics Teaching in the Secondary Schools," *School Science and Mathematics* 8 (1908): 721; Herbert G. Espy, *The Public Secondary School: A Critical Analysis of Secondary Education in the United States* (Boston: Houghton Mifflin, 1939), 241.

25. Henry Crew, "Symposium on the Purpose and Organization of Physics Teaching in the Secondary Schools," *School Science and Mathematics* 8 (1908): 723; Arthur S. Dewing, "Science Teaching in Schools: IV, Science Teaching and Induction," *School Science and Mathematics* 9 (1909): 11; W. F. Moncreiff, "A Plea for Student Laboratory Work in a First Course in Physics: Second Paper," *School Science* 3 (1904): 496.

26. H. N. Chute, "Symposium on the Purpose and Organization of Physics Teaching in the Secondary Schools," *School Science and Mathematics* 8 (1908): 726–727; Dewing, "Science Teaching in Schools: IV," 11. One of the more extended critiques that invoked the negative effects of the kindergarten movement on science instruction was given by W. D. Henderson in "The Present Status of High School Physics," *School Science and Mathematics* 8 (1908): 353–354.

27. Edwin H. Hall, "Teaching Elementary Physics," *Educational Review* 5 (1893): 327, 325.

28. Edwin H. Hall, "Natural Science Instruction (II)," *Educational Review* 30 (1905): 516; Edwin H. Hall to C. J. Bonaparte, March 17, 1898, p. 4, box 43, Records of the President of Harvard University, Charles W. Eliot, 1869–1930, Harvard University Archives.

29. Joseph Torrey Jr., *Elementary Studies in Chemistry* (New York: Henry Holt, 1899), vi. J. F. Sellers used the phrase "laboratory mania" in "A Symposium on Chemistry Requirements," *Science* 23 (1906): 734. Charles E. Bessey, "Botany by the Experimental Method," *Science* 35 (1912): 995; H. N. Chute, "The Teaching of Physics," *School Science and Mathematics* 6 (1906): 362. A general account of the turn of all science subjects from the laboratory method is provided by Espy, *The Public Secondary School*, 240–242.

30. See, for example, George R. Twiss, "Books and Literature: Present Tendencies in Science Teaching," *School and Society* 1 (1915): 387; Robert Eugene Tostberg, "Educational Ferment in Chicago, 1883–1904," PhD diss., University of Wisconsin–Madison, 1960; "Educational Chicago," *Public School Journal* 16 (1897): 541; and

Chas. W. Hargitt, "The Place and Function of Biology in Secondary Education," *Education* 25 (1905): 475, among many others.

31. Edward L. Thorndike, "Science Teaching Seen from the Outside," in *Proceedings of the Eleventh Annual Conference of the New York State Science Teachers' Association, 1906* (Albany: New York State Education Department, 1907), 71.

32. Frank Rollins, "Syllabuses and Examinations in Physics," *Educational Review* 34 (1907): 348; C. R. Mann, "The New Movement for the Reform of Physics Teaching in Germany, France, and America," in *Proceedings of the Eleventh Annual Conference of the New York State Science Teachers' Association, 1906* (Albany: New York State Education Department, 1907), 93.

33. John F. Woodhull, "Science for Culture," *School Science and Mathematics* 7 (1907): 87; George A. Works, "Results of New Movement in Teaching of Physics," *Proceedings of the Eleventh Meeting of the Central Association of Science and Mathematics Teachers* (1911), 131.

34. Charles R. Barnes, "Sciences in the High School," *School Review* 6 (1898): 647; Robert A. Millikan and Henry G. Gale, *A Laboratory Course in Physics for Secondary Schools* (Boston: Ginn, 1906), iii.

35. On the concern over specialization in the AAAS, see Sally Gregory Kohlstedt, "Creating a Forum for Science: AAAS in the Nineteenth Century," in *The Establishment of Science in America: 150 Years of the American Association for the Advancement of Science,* ed. Sally Gregory Kohlstedt, Michael M. Sokal, and Bruce V. Lewenstein (New Brunswick, NJ: Rutgers University Press, 1999), 41–48. On specialization in the sciences specifically during this time, see John Higham, "The Matrix of Specialization," and Daniel J. Kevles, "The Physics, Mathematics, and Chemistry Communities: A Comparative Analysis," both in *The Organization of Knowledge in Modern America, 1860–1920,* ed. Alexandra Oleson and John Voss (Baltimore: Johns Hopkins University Press, 1979), 4, 139–172. Another important discussion of specialization and the role disciplinary knowledge in society is found in Charles E. Rosenberg, "Toward an Ecology of Knowledge: On Discipline, Context, and History," in *No Other Gods: On Science and American Social Thought* (Baltimore: Johns Hopkins University Press, 1997), 225–239. See also Roger L. Geiger, *To Advance Knowledge: The Growth of American Research Universities, 1900–1940* (Oxford: Oxford University Press, 1986), 20–23.

36. J. H. Worst, "Danger of Overspecialization in Work in Science," *Journal of Proceedings and Addresses of the Fifty-First Annual Meeting, 1913* (Ann Arbor, MI: National Education Association, 1913), 705; John F. Woodhull, "What Specialization Has Done for Physics Teaching," *Science* 31 (1910): 729; Lewis B. Avery, "General Science in the High School," *School Science and Mathematics* 11 (1911): 740. The widening gap between research scientists and the public during the first third of the twentieth century is described in Ronald C. Tobey, *The American Ideology of National Science, 1919–1930* (Pittsburgh: University of Pittsburgh Press, 1971), 3–19.

37. C. R. Mann, "On Science Teaching, I," *School Science and Mathematics* 5 (1905): 546.

38. Hargitt, "The Place and Function of Biology in Secondary Education," 485; G. W. Myers, "The Laboratory Method in the Secondary School," *School Review* 11 (1903): 729–730. A similar equation of inductive scientific method and laboratory instruction was made by Arthur S. Dewing in "Science Teaching in Schools," *School Science and Mathematics* 8 (1908): 741–742.

39. G. W. Hunter, "The Methods, Content and Purpose of Biologic Science in the Secondary Schools of the United States," *School Science and Mathematics* 10 (1910): 105–107.

40. C. Riborg Mann, "The College Laboratory," *Education* 27 (1906): 207.

41. W. M. Smallwood, "Some Problems in Secondary Science Teaching," in *Proceedings of the Fourteenth Annual Meeting of the New York State Science Teachers' Association, 1909* (Albany: New York State Education Department, 1910), 18; C. R. Mann, "On Science Teaching VI," *School Science and Mathematics* 6 (1906): 309.

4. THE SCIENTIFIC METHOD

1. "Education at the Boston Meeting of the American Association," *Science,* n.s., 30 (1909): 871.

2. Charles W. Hargitt, "Is the Mission of Science in Education Failing?," *Education* 27 (1907): 626; "Science in the Schools Criticised by Butler," *New York Times,* December 28, 1906, 6; Elmer Ellsworth Brown, "The Outlook of the Section for Education," *Science* 27 (1908): 521–525; C. R. Mann, "Section L, Education," *Science* 27 (1908): 525. Dewey's influence was such that the final chapter of a then popular history of secondary education opened with the statement: "The keynote of current education thought seems to have been sounded by Professor John Dewey in his saying that, *the school is not preparation for life: it is life*"; Elmer Ellsworth Brown, *The Making of Our Middle Schools: An Account of the Development of Secondary Education in the United States* (New York: Longmans, Green, 1903), 436.

3. John Dewey, "Science as Subject-Matter and as Method," *Science,* n.s., 31 (1910): 121.

4. Dewey, "Science as Subject-Matter and as Method," 125.

5. C. R. Mann, "On Science Teaching, VI," *School Science and Mathematics* 6 (1906): 309.

6. L. O. Howard, "Biographical Memoir of Stephen Alfred Forbes, 1844–1930," *Biographical Memoirs,* vol. XV (Washington, DC: National Academy of Sciences, 1934), 1–54.

7. S. A. Forbes, "The Scientific Method in High School and College," *School Science* 3 (1903): 57.

8. S. A. Forbes, *The Method of Science and the Public School* (Champaign, IL: Gazette Press, 1901), 2.

9. Karl Pearson, *The Grammar of Science* (London: Walter Scott, 1892).

10. Forbes, *The Method of Science and the Public School,* 15, 16.

11. Forbes, "The Scientific Method in High School and College," 61, 63.

12. Simon Newcomb, "Art. III—Abstract Science in America, 1776–1876," *North American Review* 122 (1876): 122; Francis E. Lloyd and Maurice A. Bigelow, *The Teaching of Biology in the Secondary School* (London: Longmans, Green, 1904), 299, 301.

13. C. R. Mann, "On Science Teaching, II," *School Science and Mathematics* 5 (1905): 619, 621.

14. Alice Boardman Smuts, *Science in the Service of Children, 1893–1935* (New Haven, CT: Yale University Press, 2006), 41–44. See also Dorothy Ross, *G. Stanley Hall: The Psychologist as Prophet* (Chicago: University of Chicago Press, 1972), 341–367.

15. On functional psychology at the University of Chicago, see John M. O'Donnell, *The Origins of Behaviorism: American Psychology, 1870–1920* (New York: New York University Press, 1985), 171–177, and Andrew Feffer, *The Chicago Pragmatists and American Progressivism* (Ithaca, NY: Cornell University Press, 1993), 147–158.

16. Simon Newcomb, "The Evolution of the Scientific Investigator," *Popular Science Monthly* 66 (1904): 94. On progressive evolutionism, see Philip J. Pauly, *Controlling Life: Jacques Loeb and the Engineering Ideal in Biology* (Berkeley: University of California Press, 1990), 66–67; see also Philip J. Pauly, *Biologists and the Promise of American Life: From Meriwether Lewis to Alfred Kinsey* (Princeton, NJ: Princeton University Press, 2000), for a look at how this idea colored the biological sciences in the United States over the course of their development. The importance of the organism-environment interactions in the work of biologists at the University of Chicago during this period is described in Gregg Mitman, *The State of Nature: Ecology, Community, and American Social Thought, 1900–1950* (Chicago: University of Chicago Press, 1992), 10–47. On Dewey's work and his commitment to social activism, see Feffer, *The Chicago Pragmatists and American Progressivism,* 147–178.

17. Robert B. Westbrook, *John Dewey and American Democracy* (Ithaca, NY: Cornell University Press, 1991).

18. John Dewey, "Lecture No. 7" (First Course in Pedagogy, 1896), quoted in Arthur G. Wirth, "John Dewey's Design for American Education: An Analysis of Aspects of His Work at the University of Chicago, 1894–1904," *History of Education Quarterly* 4 (1964): 90; John Dewey, "Pedagogy: Memorandum," December 1894 [?], box 17, Presidents' Papers 1889–1925, Special Collections Research Center, University of Chicago. Data for the enrollments in the Chicago schools for the years indicated are from Department of Public Instruction, City of Chicago, *Annual Report*

of the Board of Education. A vivid description of the social and economic conditions of Chicago during these years is found in Ray Ginger, *Altgeld's America: The Lincoln Ideal versus Changing Realities* (New York: Funk and Wagnalls, 1958), 15–34.

19. John Dewey, "Lecture No. 7," 90; Dewey to Scudder Klyce, April 23, 1915 (document 03517), Correspondence of John Dewey, electronic resource, ed. Barbara Levine, Anne Sharpe, and Harriet Furst Simon, Center for Dewey Studies, Southern Illinois University at Carbondale. The curriculum of the laboratory school and the overall goals of the instruction there are described in Katherine Camp Mayhew and Anna Camp Edwards, *The Dewey School: The Laboratory School of the University of Chicago, 1896–1903* (New York: D. Appleton–Century, 1936). A nice summary of how instrumental theories of knowledge were being used to rethink the nature of schooling during this time is found in Frederick J. E. Woodbridge, "Pragmatism and Education," *Educational Review* 34 (1907): 227–240. Dewey's remarks about "man's activities" are quoted in Wirth, "John Dewey's Design for American Education," 90. On Dewey's belief in the importance of scientific thinking for social progress, see Feffer, *Chicago Pragmatists and American Progressivism,* 91–116; Daniel J. Wilson, *Science, Community, and the Transformation of American Philosophy, 1860–1930* (Chicago: University of Chicago Press, 1990), 128–130; and Westbrook, *John Dewey and American Democracy,* 138–147.

20. For a list of scientists involved in the school, see Mayhew and Edwards, *The Dewey School,* 10. The close relationship Dewey had with Loeb is described in Pauly, *Controlling Life,* 68–69. On Dewey's early intellectual growth, see Thomas C. Dalton, *Becoming John Dewey: Dilemmas of a Philosopher and Naturalist* (Bloomington: Indiana University Press, 2002). Dewey's use of science as a model for his own research as well as for the organization of the school of education at Chicago was noted in a memo to William Rainey Harper; Dewey, "Pedagogy: Memorandum," December 1894 [?], box 17, University of Chicago Presidents' Papers, 1889–1925, Department of Special Collections, University of Chicago.

21. C. R. Mann, "The Meaning of the Movement for the Reform of Science Teaching," *Educational Review* 34 (1907): 25; "Report of Meetings: Central Association of Science and Mathematics Teachers," *School Science* 3 (1904): 420–421.

22. C. M. Turton, "Central Association of Science and Mathematics Teachers," *School Science and Mathematics* 7 (1907): 66. The contents of the symposium are presented in "Symposium on the Purpose and Organization of Physics Teaching in Secondary Schools," *School Science and Mathematics* 8 (1908): 717–728, and "Symposium on the Purpose and Organization of Physics Teaching in Secondary Schools," *School Science and Mathematics* 9 (1909): 1–7; Dewey's contribution is found at 291–292. On Mann's interest in enlisting the help of educational researchers, see Mann to Michael Vincent O'Shea, September 27, 1906, and October 4, 1906, box 1, Michael Vincent O'Shea Papers, Wisconsin Historical Society Archives, Madison; C. R. Mann, "Section L, Education," 525.

23. John Dewey, *How We Think* (Boston: D. C. Heath, 1910), iii.

24. Dewey, *How We Think,* 68–78.

25. Dewey to Scudder Klyce, April 23, 1915 (document 03517), Dewey Correspondence; Dewey, *How We Think,* 79–100; Dewey, "Science as Subject-Matter and as Method," 127. By Dewey's own account, the book had little to do with science. In reply to a zoologist at Syracuse University who shared with Dewey his account of scientific method, Dewey passed along that he had written a "small book published by Heath [*How We Think*]" that touched only "somewhat on [scientific] method," Dewey to Charles C. Adams, February 15, 1916 (document 03287), Dewey Correspondence. On the importance of scientific reasoning as a model of all thought, in Dewey's view, and his effort to expand this method to all facets of human intellectual activities, see Westbrook, *John Dewey and American Democracy,* 117–149. Dewey's advocacy of a psychological interpretation of method was part of a more general turn away from intellectual formalism that was characteristic of the pragmatism of the time; see Morton White, *Social Thought in America: The Revolt against Formalism* (Boston: Beacon Press, 1957), 11–31. For a more recent account of pragmatism in American culture, see Louis Menand, *The Metaphysical Club: A Story of Ideas in America* (New York: Farrar, Straus and Giroux, 2001).

26. Dewey to Frank Manny, May 10, 1910 (document 03706), Dewey Correspondence. Publishing information (number of printings, etc.) on *How We Think* is given in the textual commentary of John Dewey, *The Middle Works, 1899–1924,* ed. Jo Ann Boydston (Carbondale: Southern Illinois University Press, 1978), 6:522–524. On the frequency of quotations, see Frank W. Scott to Dewey, October 8, 1928 (document 05910), Dewey Correspondence.

27. Frank A. Fitzpatrick, review of *How We Think* by John Dewey, *Educational Review* 40 (1910): 97.

28. Dwight W. Lott, "The Conscious Development of Scientific Ideals in Secondary Science Education," *School Science and Mathematics* 17 (1917): 417, 421; John G. Coulter, "Method in the General Science Course," *Journal of Proceedings and Addresses of the Fiftieth Annual Meeting, 1912* (Ann Arbor, MI: National Education Association, 1912), 746–747.

29. William Bishop Owen, "The Problem Method," *Chicago Schools Journal* 1, nos. 3–4 (1918): 3, 6.

30. Albert A. Michelson, "Symposium on the Purpose and Organization of Physics Teaching in Secondary Schools," *School Science and Mathematics* 9 (1909): 4; R. A. Millikan, "Present Tendencies in the Teaching of Elementary Physics," *School Science and Mathematics* 6 (1906): 121; Charles Sedgwick Minot, "The Method of Science," *Science* 33 (1911): 121.

31. Charles A. Ellwood, letter to the editor, "Scientific Method," *Science* 37 (1913): 412; Dexter S. Kimball, "Practical Work in Science Teaching," *Science* 38 (1913): 144.

32. See John M. Jordan, *Machine-Age Ideology: Social Engineering and American Liberalism, 1911–1939* (Chapel Hill: University of North Carolina Press, 1994), 1–10, 33–66; Robert H. Wiebe, *The Search for Order, 1877–1920* (New York: Hill and Wang, 1967), 164–195; David A. Hollinger, "The Problem of Pragmatism in American History," in *In the American Province: Studies in the History and Historiography of Ideas* (Baltimore: Johns Hopkins University Press, 1985), 23–43; and Lawrence A. Cremin, *The Transformation of the School: Progressivism in American Education, 1876–1957* (New York: Alfred A. Knopf, 1961), 90–126, among others.

33. B. H. Bode, review of *How We Think* by John Dewey, *School Review* 18 (1910): 642; Fitzpatrick, review of *How We Think,* 97; Max Eastman, review of *How We Think* by John Dewey, *Journal of Philosophy, Psychology, and Scientific Methods* 8 (1911): 244; W. C. Ruediger, review of *How We Think* by John Dewey, *Education* 30 (1910): 704.

34. George R. Twiss, *A Textbook in the Principles of Science Teaching* (New York: Macmillan, 1917), 21, 31.

35. Morris Meister, "The Method of the Scientists," *School Science and Mathematics* 18 (1918): 745; John Francis Woodhull, *The Teaching of Science* (New York: Macmillan, 1918), 233. See also Garfield A. Bowden, "The Project and Project Method in General Science," *School Science and Mathematics* 22 (1922): 439–446, and G. M. Ruch, "The General Science of the Future," *School Science and Mathematics* 20 (1920): 431.

5. PROBLEMS AND PROJECTS

1. "Children's Fair Date Moved to December 4–10," *New York Herald Tribune,* October 19, 1930; Ben Robertson Jr., "Modern Trends in Education and News of Interest to Teacher and Layman," *New York Herald Tribune,* November 29, 1931.

2. Adelaide Stedman, "Junior Feats in Science Spurred by Annual Fair," *New York Times,* December 3, 1933; "Tin-Can Engine Vies for Honors at Fair," *New York Times,* December 4, 1932; "Children's Fair, Opening Today, Has 481 Entries," *New York Herald Tribune,* December 4, 1932; "Children's Own Fair of Science Is Opened," *New York Times,* December 4, 1931. The best historical account of the Children's Fair and science fairs in general is found in Sevan Terzian, *Science Education and Citizenship: Fairs, Clubs, and Talent Searches for American Youth, 1918–1958* (New York: Palgrave Macmillan, 2013), 21–31. For estimates of participation and visitor attendance, see "Model Farm Shown at Children's Fair," *New York Times,* October 18, 1928; "Thousands Attend the Children's Fair," *New York Times,* October 12, 1929; "Award 143 Prizes in Children's Fair," *New York Times,* December 3, 1931; "Children Open Science Fair at History Museum," *New York Herald Tribune,* December 4, 1931. The fair started out focusing more specifically on farm, nature, and conservation topics, but quickly shifted after the second year to highlight the physical

sciences as they applied to the modern urban environment; on this point see Terzian, *Science Education and Citizenship,* 26.

3. Many exhibits, one reporter noted acerbically, were "firmly grounded in the 'scrub your teeth,' 'eat your spinach' and 'go to bed early' formula of the public schools"; "Children's Fair, Opening Today, Has 481 Entries." The seamless connection between school science instruction and science fairs and clubs was stated explicitly in the 1932 report of the Children's Science Fair: "In the flexible curriculum of the so-called 'child-centered' or progressive school, no distinction is made between the curricular and the extra-curricular" (11). The fair "can be justified by the contribution it makes to the program of science teaching. The aims or goals for the Fair should be in keeping with the best thought in psychology and educational philosophy" (29). Morris Meister, *Children's Science Fair of the American Institute: A Project in Science Education* (New York: American Institute and the American Museum of Natural History, 1932).

4. J. Stanley Brown, "The Autonomy of the High School," *Journal of Proceedings and Addresses of the Forty-Seventh Annual Meeting, 1909* (Winona, MN: National Education Association, 1909), 481.

5. G. Stanley Hall, "Some Criticism of High-School Physics, and Manual-Training and Mechanic-Arts High Schools, with Suggested Correlations," *Manual Training Magazine* 3 (1902): 189–200; Charles Riborg Mann and George Ransom Twiss, *Physics* (Chicago: Scott, Foresman, 1905), iv–vi.

6. C. R. Mann, "Physics and Daily Life," *Science* 37 (1913): 358.

7. Charles E. Dull, *Essentials of Modern Physics* (New York: Henry Holt, 1922); Henry S. Carhart and Horatio N. Chute, *Practical Physics,* rev. ed. (Boston: Allyn and Bacon, 1927). Millikan and Gale's *Practical Physics* followed the same pattern as the others; Robert A. Millikan, Henry Gordon Gale, and Willard R. Pyle, *Practical Physics,* rev. ed. (New York: Ginn, 1922). See also Charles Gilpin Cook, *A Practical Chemistry for High School Students* (New York: D. Appleton, 1916), and William McPherson and William Edwards Henderson, *Chemistry and Its Uses: A Textbook for Secondary Schools* (New York: Ginn, 1922).

8. Carhart and Chute, *Practical Physics,* 214; Dull, *Essentials of Modern Physics,* 218; Oscar M. Stewart, Burton L. Cushing, and Judson R. Towne, *Physics for Secondary Schools* (Boston: Ginn, 1932), 105, 449.

9. Henry R. Linville, "Old and New Ideals in Biology Teaching," *School Science and Mathematics* 10 (1910): 210. On the rise of biology, see Philip J. Pauly, "The Development of High School Biology: New York City, 1900–1925," *Isis* 82 (1991): 662–668. See also Philip J. Pauly, *Biologists and the Promise of American Life: From Meriwether Lewis to Alfred Kinsey* (Princeton, NJ: Princeton University Press, 2000) for a discussion of how biologists saw high school teaching as part of a broader project of public cultivation during this period. On the temperance movement in American schools, see Jonathan Zimmerman, *Distilling Democracy: Alcohol Educa-*

tion in America's Public Schools, 1880–1925 (Lawrence: University Press of Kansas, 1999).

10. Linville, "Old and New Ideals in Biology Teaching," 212.

11. Linville, "Old and New Ideals in Biology Teaching," 212–213 (emphasis in original). See also Harold B. Shinn, "Biology in the High School of Tomorrow," *School Science and Mathematics* 18 (1918): 495–499.

12. George W. Hunter, *A Civic Biology, Presented in Problems* (New York: American Book Company, 1914), 7; Hunter, *Civic Biology,* 19. For examples of textbooks that placed less emphasis on student problems, see Truman J. Moon, *Biology for Beginners* (New York: Henry Holt, 1921); and W. M. Smallwood, Ida L. Reveley, and Guy A. Bailey, *New Biology* (Boston: Allyn and Bacon, 1924).

13. Frederick L. Fitzpatrick and Ralph E. Horton, *Biology* (Boston: Houghton Mifflin, 1937), 181; George W. Hunter, *Problems in Biology* (New York: American Book Company, 1939), 13.

14. Fletcher B. Dresslar, "A Brief Survey of Educational Progress during the Decade 1900 to 1910," *Report of the Commissioner of Education for the Year Ended June 30, 1911* (Washington, DC: Government Printing Office, 1912), 9–10.

15. Harold B. Shinn, "The Movement toward a Unified Science Course in Secondary Schools," *School Science and Mathematics* 14 (1914): 779. For concerns about enrollment, see, for example, W. D. Lewis, "What Is the Matter with Our High School Science Courses?," *Proceedings of the Fourteenth Annual Meeting of the New York State Science Teachers' Association, 1909* (Albany: University of the State of New York, 1910), 26–36; Albert C. Herre, "'Detested Science' in the High School: A Reply to Percy E. Rowell," *School Science and Mathematics* 11 (1911); 169; Willard J. Fisher, "Is Science Really Unpopular in High Schools?" *Science* 35 (1912): 94–98; Willard J. Fisher, "The Drift in Secondary Education," *Science* 36 (1912): 587–590; Elliot R. Downing, "The Scientific Trend in Secondary Schools," *Science* 41 (1915): 232–235; Elliot R. Downing, "Enrollment in Science in the High Schools," *Science* 46 (1917): 351–352. These authors echoed concerns found in earlier articles; see, for example, G. C. Bush, "The Status of the Physical Sciences in the High School," *School Science and Mathematics* 5 (1905): 431–436.

16. R. O. Austin, "The Need and Scope of a First Year General Science Course," *School Science and Mathematics* 11 (1911): 217; Charles Emerson Peet, "What Shall the First-Year High-School Science Be?," *Journal of Proceedings and Addresses of the Forty-Seventh Annual Meeting, 1909* (Winona, MN: National Education Association, 1909), 811. For the history of the early courses, see Austin, "The Need and Scope of a First Year General Science Course," and Otis W. Caldwell, "The Course in General Elementary Science for the First Year of the High School," *Proceedings of the Ninth Meeting of the Central Association of Science and Mathematics Teachers* (1909): 115–127.

17. Clarence M. Pruitt, "William Lewis Eikenberry," *Science Education* 40, no. 4 (1956): 259–261.

18. Otis W. Caldwell, "Science and the Passing Years," outline of an address given May 8, 1942, in honor of W. L. Eikenberry, unpublished typescript, Otis W. Caldwell Papers (private papers in the possession of John Heffron, Soka International University, Aliso Viejo, CA; hereafter cited as Caldwell Papers). My sincere thanks to Prof. Heffron for sharing these materials with me. The details on Caldwell's life can be found scattered among the meeting minutes of the Central Association of Science and Mathematics Teachers as well as in John Marcher Heffron, "Science, Southerness, and Vocationalism: Rockefeller's 'Comprehensive System' and the Reorganization of Secondary School Science Education, 1900–1920," PhD diss., University of Rochester, 1988. Otis W. Caldwell to M. M. Williams, January 10, 1941, Caldwell Papers.

19. William L. Eikenberry, "The General-Science Course in the University High School," *School Review* 20 (1912): 219; Otis W. Caldwell, "The Course in General Elementary Science for the First Year of the High School," *Proceedings of the Ninth Meeting of the Central Association of Science and Mathematics Teachers* (1909), 117; Otis W. Caldwell to M. M. Williams, January 10, 1941; Otis W. Caldwell, "Science and the Passing Years," 1942, Caldwell Papers.

20. Eikenberry, "The General-Science Course in the University High School," 221, 222; Caldwell, "The Course in General Elementary Science for the First Year of the High School," 118.

21. Otis W. Caldwell and William L. Eikenberry, *Elements of General Science* (Boston: Ginn, 1914); Otis W. Caldwell, W. L. Eikenberry, and Charles J. Pieper, *A Laboratory Manual for Work in General Science* (Boston: Ginn, 1915), 85.

22. Notable examples include Bertha M. Clark, *General Science* (New York: American Book Company, 1912); Percy E. Rowell, *Introduction to General Science with Experiments* (New York: Macmillan, 1911); John C. Hessler, *The First Year of Science* (Chicago: Benj. H. Sanborn, 1914); Frederic Delos Barber, Merton Leonard Fuller, John Lossen Pricer, and Howard William Adams, *First Course in General Science* (New York: Henry Holt, 1916); Lewis Elhuff, *General Science: First Course* (Boston: D. C. Heath, 1916); William H. Snyder, *General Science* (Boston: Allyn and Bacon, 1925); and Ralph K. Watkins and Ralph C. Bedell, *General Science for Today* (New York: Macmillan, 1932).

23. Ginn and Company, advertisement, *General Science Quarterly* 1 (1916): 63 (emphasis in original).

24. Abstract of an address by John Dewey entitled "Disciplinary Value of Science Teaching," in "Reports of Meetings: Central Association of Science and Mathematics Teachers," *School Science* 3 (1904): 420–421; John Dewey, "Method in Science Teaching," *General Science Quarterly* 1 (1916): 3.

25. Eikenberry, *The Teaching of General Science,* 70, 76–77.

26. "General Science at the National Education Association in New York," *General Science Quarterly* 1 (1916): 58; "Organize a General Science Club," *General Sci-*

ence Quarterly 1 (1916): 59; "General Science Club of New England," *General Science Quarterly* 1 (1916): 62.

27. John M. Coulter, "The Mission of Science in Education," *School Review* 23 (1915): 8 (emphasis in original); H. L. Wieman, "Teaching Scientific Method versus Teaching the Facts of Science," *School and Society* 3 (1916): 853. Similar objections are found in J. G. Coulter, "A Four-Year Course in Science in the High School," *School and Society* 1 (1915): 226–234, and John G. Coulter, "Proposed Status of Science Instruction in the Junior-Senior High School Organization," *Educational Administration and Supervision* 1 (1915): 639–645.

28. Wieman, "Teaching Scientific Method versus Teaching the Facts of Science," 854; R. A. Millikan, "The Problem of Science Teaching in the Secondary Schools," *School Science and Mathematics* 25 (1925): 972.

29. Ailsie M. Heineman, "A Study of General Science Textbooks," *General Science Quarterly* 13 (1928): 11; W. R. Leker, "The Articulation of General Science with the Special Sciences," *School Science and Mathematics* 25 (1925): 730; Frank M. Phillips, *Statistics of Public High Schools, 1927–1928,* Bulletin No. 35, Office of Education, U.S. Department of the Interior (Washington, DC: Government Printing Office, 1929); A. J. Tieje, "A Course in General Science," *Science* 58 (1923): 279.

30. Some textbooks that made the connection between general science and the project method include Garfield A. Bowden, *General Science with Experimental and Project Studies* (Philadelphia: P. Blakiston's Son, 1923); Edgar A. Bedford, *General Science: A Book of Projects* (Boston: Allyn and Bacon, 1921); and Lewis Elhuff, *New Laboratory Manual: A Book of Projects for General Science* (Boston: D. C. Heath, 1921). Articles in the science education literature touting the connection include George D. von Hofe Jr., "General Science Is Project Science," *School Science and Mathematics* 15 (1915): 751–757; William Sayles Wake, "The Project in General Science," *School Science and Mathematics* 19 (1919): 643–650; and G. H. Trafton, "Project Teaching in General Science," *School Science and Mathematics* 21 (1921): 315–322.

31. General Education Board, *The General Education Board: An Account of Its Activities, 1902–1914* (New York: General Education Board, 1915), 61, 70. For a history of the term "project" in education, see Nelson L. Bossing, *Progressive Methods of Teaching in Secondary Schools* (Boston: Houghton Mifflin, 1935), 485–507. For background on the country life movement, see Amrys O. Williams, "Cultivating Modern America: 4-H Clubs and Rural Development in the Twentieth Century," PhD diss., University of Wisconsin–Madison, 2012, 24–46, and Massachusetts Board of Education, *Agricultural Project Study: Information and Suggestions for School Officers and Instructors as to Courses and Methods of Agricultural Project,* Bulletin No. 4 (Boston: Massachusetts State Board of Publication, 1912).

32. C. R. Mann, "What Is Industrial Science?," *Science* 39 (1914): 522, 523. References to corn clubs are also found in J. A. Randall, "Project Teaching," *Journal of Proceedings and Addresses of the Fifty-Third Annual Meeting, 1915* (Ann Arbor,

MI: National Education Association, 1915), 1009, and John F. Woodhull, "General Science—Summary of Opinions under Revision," *School Science and Mathematics* 14 (1914): 600 (emphasis in original).

33. The first example is from George W. Hunter and Walter G. Whitman, *Civic Science in the Home* (New York: American Book Company, 1921), 20; the second is from Edgar F. Van Buskirk and Edith Lillian Smith, *The Science of Everyday Life* (Boston: Houghton Mifflin, 1919), 93.

34. C. R. Mann, "Project Teaching," *General Science Quarterly* 1 (1916): 14; J. C. Moore, "Projects," *General Science Quarterly* 1 (1916): 15. In 1923, Ralph Watkins wrote that "most of the proponents of the project idea are direct or second generation students of Dewey, and it is not difficult to trace the idea back to its source"; Ralph K. Watkins, "The Technique and Value of Project Teaching in General Science," *General Science Quarterly* 7 (1923): 244.

35. W. F. Roecker, "An Elementary Course in General Science: Content and Method," *School Science and Mathematics* 14 (1914): 757; Mendel E. Branom, *The Project Method in Education* (Boston: Richard G. Badger, 1919), 60, 13.

36. See Ada L. Weckel, "Are Any Principles of Organization of General Science Evidenced by the Present Textbooks in the Subjects?," *General Science Quarterly* 6 (1922): 386–395; Will S. Kellogg, "A Survey of the Status of General Science in California," *General Science Quarterly* 6 (1922): 373–383; and Warren W. Knox, "Ninth Grade Science," *Teaching Biologist* 3, no. 7 (1934): 1. On the rise of traditional environmental science courses, see Stephen A. Laubach, "From Civic Conservation to the Age of Ecology: The Rise and Synthesis of Ecological Ideas in the American High School Science Curriculum, 1900–1980," PhD diss., University of Wisconsin–Madison, 2013.

37. Morris Meister, "The Method of the Scientists," *School Science and Mathematics* 18 (1918): 744–745.

38. Kilpatrick, "The Project Method," 320, 323, 332–333.

39. Herbert M. Kliebard, *The Struggle for the American Curriculum, 1893–1958*, 3rd ed. (New York: RoutledgeFalmer, 2004), 135–140. One teacher educator wrote that the creation of a display or exhibit was the final step in the project method; J. O. Frank, *How to Teach General Science: Notes and Suggestions of Practical Aid to Every General Science Teacher* (Philadelphia: P. Blakiston's Son, 1926), 94–95.

40. George W. Hunter, *Science Teaching at Junior and Senior High School Levels* (New York: American Book Company, 1934), 488, 496; Carleton E. Preston, *The High School Science Teacher and His Work* (New York: McGraw-Hill, 1936), 189; "School Laboratories Called Over-Equipped," *New York Times*, August 12, 1929; Elwood D. Heiss, Ellsworth S. Obourn, and Charles W. Hoffman, *Modern Science Teaching* (New York: Macmillan, 1950), 247–269; Hunter, *Science Teaching at Junior and Senior High School Levels*, 503.

41. Paul B. Mann, "Is It Worth While to Teach Science?," *Teaching Biologist* 7 (1938): 68. The full articulation of this new approach to education was laid out in a

National Education Association–sponsored report; Department of the Interior, Bureau of Education, *Cardinal Principles of Secondary Education: A Report of the Commission on the Reorganization of Secondary Education, Appointed by the National Education Association,* Bulletin No. 35 (Washington, DC: Government Printing Office). The best discussion of the history of the *Cardinal Principles* report is found in Kliebard, *Struggle for the American Curriculum.*

6. THE WAR ON METHOD

1. George Wald, Chair of the Subcommittee on Science and Mathematics, "Proposals for a Harvard Program," box 3, Records of the Committee on General Education in a Free Society, Harvard University Archives; Committee on the Objectives of General Education in a Free Society, *General Education in a Free Society: Report of the Harvard Committee* (Cambridge, MA: Harvard University Press, 1945), 158. Among the correspondence from Conant to the Committee are J. B. Conant to Paul H. Buck, July 18, 1944, and Conant memorandum to Dean Buck, December 6, 1944, box 259, Records of the President of Harvard University— James Bryant Conant, Harvard University Archives (hereafter Conant Presidential Records).

2. Vannevar Bush, "Planning in Science," in *Science and Life in the World: The George Westinghouse Centennial Forum, May 16, 17, 18, 1946,* vol. 1, *Science and Civilization* (New York: McGraw-Hill, 1946), 54.

3. Daniel J. Kevles, *The Physicists: The History of a Scientific Community in Modern America* (Cambridge, MA: Harvard University Press, 1995), 302–323. On the social and cultural impact of the bomb, see Paul S. Boyer, *By the Bomb's Early Light: American Thought and Culture at the Dawn of the Atomic Age* (Chapel Hill: University of North Carolina Press, 1985). Stuart W. Leslie, *The Cold War and American Science: The Military-Industrial-Academic Complex at MIT and Stanford* (New York: Columbia University Press, 1993); Rebecca S. Lowen, *Creating the Cold War University: The Transformation of Stanford* (Berkeley: University of California Press, 1997); Peter Galison and Bruce Hevly, eds., *Big Science: The Growth of Large-Scale Research* (Stanford, CA: Stanford University Press, 1992).

4. David A. Hollinger, "Free Enterprise and Free Inquiry: The Emergence of Laissez-Faire Communitarianism in the Ideology of Science in the United States," *New Literary History* 21 (1990): 897–919; Andrew Jewett, *Science, Democracy, and the American University: From the Civil War to the Cold War* (New York: Cambridge University Press, 2012), 302–334. See also John L. Rudolph, *Scientists in the Classroom: The Cold War Reconstruction of American Science Education* (New York: Palgrave Macmillan, 2002), 33–55.

5. The marginalization of the social sciences after the war is discussed by Jewett, *Science, Democracy, and the American University,* 310–320.

6. Simon Newcomb, "Abstract Science in America, 1776–1876," *North American Review* 122 (Jan. 1876): 122–123; Karl Pearson, *The Grammar of Science,* 2nd ed. (London: Adam and Charles Black, 1900), 7, 9; John Dewey, "The Supreme Intellectual Obligation," *Science Education* 18 (1934): 1–4.

7. F. Stuart Chapin, "Progress in Methods of Inquiry and Research in the Social and Economic Sciences," *Scientific Monthly* 19 (1924): 390–399; Odell Hauser, "Science Aims to End Prejudice in Politics," *New York Times,* October 11, 1925, 205; "Scientists Urged to Cure Ill World," *New York Times,* December 29, 1938, 9. On the general tendencies in the social sciences, see Jewett, *Science, Democracy, and the American University,* 272–301, and Peter J. Kuznick, *Beyond the Laboratory: Scientists as Political Activists in 1930s America* (Chicago: University of Chicago Press, 1987), 38–64.

8. "Measuring Talent: Scientific Methods Now Being Used in Educational Centers," *Los Angeles Times,* August 20, 1922, III-18; "Urges Facts on Fashions," *New York Times,* August 7, 1928, 37; "Finds Proof of Christ in Scientific Method," *New York Times,* December 26, 1932, 21.

9. "Millikan Sees World Remade by Use of Scientific Method," *New York Times,* December 21, 1930, 118; "'Biology Rather than Physics Will Bring the Big Changes'—Also, Says Dr. Millikan, the Scientific Method Will Aid in Government," *New York Times,* September 13, 1931, XX3; "Expert Leadership Urged: Millikan Says Presidents and Governors Should Be Scientific in Making Appointments," *Los Angeles Times,* November 13, 1934, A1. For more on Millikan's views on the relationship between science and society, see Robert H. Kargon, *The Rise of Robert Millikan: Portrait of a Life in American Science* (Ithaca, NY: Cornell University Press, 1982); "World Gets Warning: Drift toward Chaos Seen," *Los Angeles Times,* June 23, 1934, 1.

10. Paul Forman, "Behind Quantum Electronics: National Security as Basis for Physical Research in the United States, 1940–1960," *Historical Studies in the Physical and Biological Sciences* 18 (1987): 152; A. Hunter Dupree, *Science in the Federal Government: A History of Policies and Activities to 1940* (Cambridge, MA: Harvard University Press, 1957).

11. Zacharias quoted in Forman, "Behind Quantum Electronics," 152. On the connections between science and the government, see Silvan S. Schweber, "The Mutual Embrace of Science and the Military: ONR and the Growth of Physics in the United States after World War II," in *Science, Technology, and the Military,* ed. Everett Mendelsohn, Merritt Roe Smith, and Peter Weingart (Dordrecht: Kluwer Academic, 1988), 3–45; see also Roger L. Geiger, "Science, Universities, and National Defense, 1945–1970," *Osiris* 7 (1992): 26–48; and Kenneth M. Jones, "The Government-Science Complex," in *Reshaping America: Society and Institutions, 1945–1960,* ed. Robert H. Bremner and Gary W. Reichard (Columbus: Ohio State University Press, 1982), 315–342; and Rudolph, *Scientists in the Classroom,* 33–38.

12. Rudolph, *Scientists in the Classroom,* 38–52; quotation from Melba Phillips, "Dangers Confronting American Science," *Science* 116 (1952): 440.

13. Vannevar Bush, *Science: The Endless Frontier: A Report to the President* (Washington, DC: United States Government Printing Office, 1945), 33; Daniel J. Kevles, "The National Science Foundation and the Debate over Postwar Research Policy, 1942–1945: A Political Interpretation of *Science—The Endless Frontier*," *Isis* 68 (1977): 5–26; Daniel Lee Kleinman, *Politics on the Endless Frontier: Postwar Research Policy in the United States* (Durham, NC: Duke University Press, 1995).

14. Bush, "Planning in Science," 54; Hollinger, "Free Enterprise and Free Inquiry," 897–919; Kevles, "The National Science Foundation and the Debate over Postwar Research Policy," 5–26. Quotation about "majority" from P. W. Bridgman, "The Prospect for Intelligence," *Yale Review* 34 (1945): 458.

15. Scates quoted in Thomas F. Gieryn, *Cultural Boundaries of Science: Credibility on the Line* (Chicago: University of Chicago Press, 1999), 80, 65–84. Mark Solovey, *Shaky Foundations: The Politics-Patronage-Social Science Nexus in Cold War America* (New Brunswick, NJ: Rutgers University Press).

16. Daniel P. Thurs, "Scientific Methods," in *Wrestling with Nature: From Omens to Science*, ed. Peter Harrison, Ronald L. Numbers, and Michael H. Shank (Chicago: University of Chicago Press, 2011), 325–330.

17. Otis W. Caldwell and Francis D. Curtis, *Science for Today* (Boston: Ginn, 1936), 7; Truman J. Moon, Paul B. Mann, and James H. Otto, *Modern Biology*, rev. ed. (New York: Henry Holt, 1951), 8 (emphasis in original); Caldwell and Curtis, *Science for Today*, 9.

18. George W. Hunter, *Science Teaching at Junior and Senior High School Levels* (New York: American Book Company, 1934), 213. Hunter gained a measure of notoriety as the author of *Civic Biology*, which was the biology textbook at the center of the Scopes trial in 1925; see Adam R. Shapiro, *Trying Biology: The Scopes Trial, Textbooks, and the Antievolution Movement in American Schools* (Chicago: University of Chicago Press, 2013); Elwood D. Heiss, Ellsworth S. Obourn, and C. Wesley Hoffman, *Modern Methods and Materials for Teaching Science* (New York: Macmillan, 1940), 16.

19. Nelson B. Henry, ed., *The Forty-Sixth Yearbook of the National Society of the Study of Education: Part 1: Science Education in American Schools* (Chicago: University of Chicago Press, 1947), 15. Similar ideas about the control of scientific research were expressed in Heiss, Obourn, and Hoffman, *Modern Methods and Materials for Teaching Science*, 4–5.

20. Arthur Zilversmit, *Changing Schools: Progressive Education Theory and Practice, 1930–1960* (Chicago: University of Chicago Press, 1993), 91; "Nation Tackles School Crisis," *Life*, September 26, 1955, 31–37; Diane Ravitch, *The Troubled Crusade: American Education, 1945–1980* (New York: Basic Books, 1983), 6; "Back to School—and a Bigger Jam," *U.S. News and World Report*, September 4, 1953, 22–23.

21. C. R. Mann, "Project Teaching," *General Science Quarterly* 1 (1916): 14; Elliot R. Downing, "The Elements and Safeguards of Scientific Thinking," *Scientific*

Monthly 26 (1928): 231; Oreon Keeslar, "The Elements of Scientific Method," *Science Education* 29 (1945): 277.

22. Keeslar, "The Elements of Scientific Method," 277; Hunter, *Science Teaching at Junior and Senior High School Levels,* 227; Downing, "The Elements and Safeguards of Scientific Thinking," 231. The emphasis on training through repeated use is also found in M. Louise Nichols, "The High School Student and Scientific Method," *Journal of Educational Psychology* 20 (1929): 196–204; and Elwood D. Heiss, Ellsworth S. Obourn, and C. Wesley Hoffman, *Modern Science Teaching: A Revision of Modern Methods and Materials for Teaching Science* (New York: Macmillan, 1950), 96.

23. Neil E. Stevens, "The Moral Obligation to Be Intelligible," *Scientific Monthly* 70 (1950): 115; Harold K. Schilling, "A Human Enterprise," *Science* 127 (1958): 1324; J. T., "Is There a Scientific Method?," *Science* 126 (1957): 431; C. D. Darlington, *The Conflict of Science and Society* (London: Watts, 1948), 6.

24. P. W. Bridgman, "On 'Scientific Method,'" *Teaching Scientist* 6, no. 2 (1949): 23; Bridgman, "The Prospect for Intelligence," 450. The embrace of creativity and intuition in scientific thought is treated in Jamie Cohen-Cole, *The Open Mind: Cold War Politics and the Sciences of Human Nature* (Chicago: University of Chicago Press, 2014), and Jewett, *Science, Democracy, and the American University,* 318–320.

25. Conant not only had initiated the Redbook project but regularly attended committee meetings and participated in shaping its vision. See, for example, James B. Conant, Memorandum to Dean Buck, December 6, 1944, box 259, Conant Presidential Records; Record of Interview: DCJ, CD and James B. Conant, Subject: Proposed Science Course, May 14, 1947, box 151, Carnegie Corporation of New York Records, Rare Book and Manuscript Library, Columbia University (hereafter CCNY Records); see also Christopher Hamlin, "The Pedagogical Roots of the History of Science: Revisiting the Vision of James Bryant Conant," *Isis* 107 (2016): 282–308.

26. Remarks by Conant on receiving the Arches of Science Award in Seattle, October 24, 1967, box 17, folder "On Understanding Science," Papers of James Bryant Conant, Harvard University Archives (hereafter Conant Papers); James G. Hershberg, *James B. Conant: Harvard to Hiroshima and the Making of the Nuclear Age* (New York: Knopf, 1993).

27. J. B. Conant to Dean Paul H. Buck, memorandum, "Prospectus of a Course on 'The Advancement of Knowledge in Modern Times,'" July 18, 1944, box 259, Conant Presidential Records.

28. James Bryant Conant, *Education in a Divided World: The Function of the Public Schools in Our Unique Society* (Cambridge, MA: Harvard University Press, 1948), 118–119.

29. Conant, *Education in a Divided World,* 121.

30. James B. Conant, *On Understanding Science: An Historical Approach* (New Haven, CT: Yale University Press, 1947), 13; Conant, *On Understanding Science,* 18;

James Bryant Conant and Leonard K. Nash, eds., *Harvard Case Histories in Experimental Science,* vol. 1 (Cambridge, MA: Harvard University Press, 1964).

31. Conant, *On Understanding Science,* 16, 98–109, 14–15.

32. Fletcher G. Watson, P. LeCorbeiller, Edwin C. Kemble, I. Bernard Cohen, and Edward S. Castle, "Science in the General Education Program at Harvard University," in *Science in General Education,* ed. Earl J. McGrath (Dubuque, IA: Wm. C. Brown, 1948), 103–104; I. Bernard Cohen and Fletcher G. Watson, eds., *General Education in Science* (Cambridge, MA: Harvard University Press, 1952), vi–ix; James B. Conant to Chester [Charles] Dollard, June 5, 1950, box 161, CCNY Records; Thomas S. Kuhn, *The Structure of Scientific Revolutions* (Chicago: University of Chicago Press, 1962). On Kuhn, see Steve Fuller, *Thomas Kuhn: A Philosophical History for Our Times* (Chicago: University of Chicago Press, 2000). Kuhn's book laid out a view of science that was complex and differed radically from past accounts of what science was and how it operated; in no small measure, it was this view that served as the foundation of the new field of science studies. See Hamlin, "The Pedagogical Roots of the History of Science," 282–308; Fuller, *Thomas Kuhn,* 179–226. On the impact of *Structure,* see Robert J. Richards and Lorraine Daston, eds., *Kuhn's Structure of Scientific Revolutions at Fifty: Reflections on a Science Classic* (Chicago: University of Chicago Press, 2016); John H. Zammito, *A Nice Derangement of Epistemes: Post-Positivism in the Study of Science from Quine to Latour* (Chicago: University of Chicago Press, 2004), 52–150; and Fuller, *Thomas Kuhn,* 318–331.

33. "Nash Wounded in Experiment," *Harvard Crimson,* November 4, 1954; George E. Erikson, "The General Education Course in Biology: Laboratory Work and General Objectives," in *General Education in Science,* ed. I. Bernard Cohen and Fletcher G. Watson (Cambridge, MA: Harvard University Press, 1952), 176.

34. Attendees at the 1949 conference are listed in Record of Interview: FA and CD, Subject: Harvard University Conference on the Place of Science in General Education, July 9 and 10, 1949, box 161, CCNY Records; those presenting papers at the 1950 summer conference are listed in Cohen and Watson, eds., *General Education in Science;* for the 1951 conference, see Report of Conference on Science in General Education, Cambridge, Massachusetts, July 9–12, 1951, box 161, CCNY Records; Hamlin, "The Pedagogical Roots of the History of Science," 282–308.

35. See, for example, Cohen and Watson, eds., *General Education in Science,* 13, 30, 56, 84, 98, 99, 117, 178; I. B. Cohen, "The History of Science and the Teaching of Science," in *General Education in Science,* ed. Cohen and Watson, 85.

36. W. Y. Elliott to Charles Dollard, November 21, 1951, box 161, CCNY Records; press release, "President James B. Conant Will Offer a Course for High School Science Teachers in the Harvard Summer School of 1952," February 10, 1952, box 443, Conant Presidential Records; James B. Conant to W. Y. Elliott, February 20, 1952, box 443, Conant Presidential Records; Report of the Science Education Program—1952, Harvard Summer School, January 1953, box 471, Conant Presidential

Records; James Conant to Charles Dollard, October 8, 1952, box 161, CCNY Records.

37. James B. Conant, *Modern Science and Modern Man* (New York: Columbia University Press, 1952), 19–26; James B. Conant to John Dewey, February 27, 1951, Conant Papers. Dewey acknowledged Conant's letter but was convalescing in Arizona at the time (the book was sent to Dewey's home in New York). Dewey died about a year later. There's no record of Dewey's thoughts on Conant's chapter.

38. John Dewey, *How We Think: A Restatement of the Relation of Reflective Thinking to the Educative Process* (Boston: D. C. Heath, 1933), 115–116; William Kent to James B. Conant, June 15, 1952, box 18, Conant Papers; James B. Conant to John Dewey, February 27, 1951, box 41, Conant Presidential Records; James B. Conant to William Kent, July 22, 1952, box 18, Conant Papers.

39. Philip Morrison, "Dr. Conant Considers the Tactics, Strategy and Goal of Science," *New York Times,* April 27, 1947; Joseph T. Clark, review of *Science and Common Sense, America* 85 (July 21, 1951): 400; Jane Oppenheimer, "On Understanding President Conant," review of *Science and Common Sense* by James B. Conant, *Quarterly Review of Biology* 26 (1951): 364; Vannevar Bush, "What Every Layman Should Know," review of *Science and Common Sense* by James B. Conant, *Saturday Review,* February 17, 1951, 14–15.

40. Leo E. Klopfer and Fletcher G. Watson, "Historical Materials and High School Science Teaching," *Science Teacher* 24 (1957): 264–265, 292–293; Leopold E. Klopfer and William W. Cooley, "The *History of Science Cases for High Schools* in the Development of Student Understanding of Science and Scientists," *Journal of Research in Science Teaching* 1 (1963): 33–47. Accounts of this shift toward scientist involvement are given by Diane Ravitch, *The Troubled Crusade: American Education, 1945–1980* (New York: Basic Books, 1983), 228–266. Joseph J. Schwab described the shift as well at the time in his "The Teaching of Science as Enquiry," in *The Teaching of Science,* ed. Joseph J. Schwab and Paul F. Brandwein (Cambridge, MA: Harvard University Press, 1962), 21–24; see also William E. Brownson and Joseph J. Schwab, "American Science Textbooks and Their Authors, 1915 and 1955," *School Review* 71 (1963): 170–180.

7. ORIGINS OF INQUIRY

1. "Reds Fire 'Moon' into Sky," *Chicago Daily Tribune,* October 5, 1957, 1; "Russ Satellite Circling Earth," *Los Angeles Times,* October 5, 1957, 1; Paul Dickson, *Sputnik: The Shock of the Century* (New York: Walker, 2001); Robert A. Divine, *The Sputnik Challenge: Eisenhower's Response to the Soviet Satellite* (New York: Oxford University Press, 1993), 15.

2. James T. Patterson, *Grand Expectations: The United States, 1945–1974* (New York: Oxford University Press, 1996), 165–173; Paul Boyer, *By the Bomb's Early*

Light: American Thought and Culture at the Dawn of the Atomic Age (Chapel Hill: University of North Carolina Press, 1994), 336–337; Stephen J. Whitfield, *The Culture of the Cold War* (Baltimore: Johns Hopkins University Press, 1991), 1–25; Ellen Schrecker, *Many Are the Crimes: McCarthyism in America* (Boston: Little, Brown, 1998), 154–200; Audra J. Wolfe, *Competing with the Soviets: Science, Technology, and the State in Cold War America* (Baltimore: Johns Hopkins University Press, 2013). On the effect of Sputnik, see Divine, *The Sputnik Challenge.*

3. Diane Ravitch, *The Troubled Crusade: American Education, 1945–1980* (New York: Basic Books, 1983), 228–266; Herbert M. Kliebard, *The Struggle for the American Curriculum, 1893–1958,* 3rd ed. (New York: RoutledgeFalmer, 2004), 250–270. Notable attacks on public schools at the time included Mortimer Smith, *And Madly Teach: A Layman Looks at Public School Education* (Chicago: Henry Regnery, 1949); Bernard Iddings Bell, *Crisis in Education: A Challenge to American Complacency* (New York: Whittlesey House, 1949); Arthur E. Bestor, *Educational Wastelands: The Retreat from Learning in Our Public Schools* (Urbana: University of Illinois Press, 1953); and Albert Lynd, *Quackery in the Public Schools* (Boston: Little, Brown, 1953).

4. "Sputnik Congress Opens!," *Chicago Daily Tribune,* January 8, 1958, 1; "54 Chicagoans Tell Ways to Counter Russian Sputnik," *Chicago Daily Tribune,* January 8, 1958, 10. The Nobel laureates were Hans Bethe, Hermann Muller, Edward Purcell, I. I. Rabi, and Glenn Seaborg.

5. P. W. Bridgman, "The Prospect for Intelligence," *Yale Review* 34 (1945): 458.

6. John L. Rudolph, "Myth 23. That the Soviet Launch of *Sputnik* Caused the Revamping of American Science Education," in *Newton's Apple and Other Myths about Science,* ed. Ronald L. Numbers and Kostas Kampourakis (Cambridge, MA: Harvard University Press, 2015), 186–192.

7. John L. Rudolph, *Scientists in the Classroom: The Cold War Reconstruction of American Science Education* (New York: Palgrave Macmillan, 2002), 83.

8. AIBS Executive Committee meeting minutes, January 4, 1952, box 2, Records of the American Institute of Biological Sciences, Washington, DC; National Research Council, Conference on Biological Education meeting minutes, March 10, 1953; CEP meeting minutes, November 13, 1954, Central Policy Files, Biology and Agriculture / Committee on Educational Policies, National Academy of Sciences Archives, Washington, DC (hereafter NAS / CEP); see also Richard E. Paulson, "The Committee on Educational Policies," *AIBS Bulletin* 5, no. 5 (1955): 17–18; Frank L. Campbell, "Committee on Educational Policies in Biology," 1958, NAS / CEP; Chester A. Lawson and Richard E. Paulson, *Laboratory and Field Studies in Biology: A Sourcebook for Secondary Schools* (New York: Holt, Rinehart and Winston, 1960); Paul Westmeyer, "The Chemical Bond Approach to Introductory Chemistry," *School Science and Mathematics* 61 (1961): 317.

9. Divine, *The Sputnik Challenge;* James R. Killian, *Sputnik, Scientists, and Eisenhower: A Memoir of the First Special Assistant to the President for Science and*

Technology (Cambridge, MA: MIT Press, 1977); Nicholas DeWitt, *Soviet Professional Manpower: Its Education, Training, and Supply* (Washington, DC: National Science Foundation, 1955). On the response of Congress to the report on Soviet education, see Rudolph, *Scientists in the Classroom,* 74–77; U.S. House Committee on Science and Technology, Subcommittee on Science, Research, and Technology, *The National Science Foundation and Pre-College Science Education: 1950–1975,* report prepared by Science Policy Research Division, 94th Cong., 2d sess., 1976, Committee Print 61-660. On the specific funding of biology, chemistry, and physics, see Appendix C, 230. On the legislative history of the NDEA, see Barbara Barksdale Clowse, *Brainpower for the Cold War: The Sputnik Crisis and National Defense Education Act of 1958* (Westport, CT: Greenwood Press, 1981). For the history of the development of the physics and biology projects, see Rudolph, *Scientists in the Classroom.* The details of the chemistry projects can be found in Laurence E. Strong, "Chemistry as a Science in the High School," *School Review* 70 (1962): 44–50, and Richard J. Merrill and David W. Ridgway, *The CHEM Study Story* (San Francisco: W. H. Freeman, 1969). See also Peter B. Dow, *Schoolhouse Politics: Lessons from the Sputnik Era* (Cambridge, MA: Harvard University Press, 1991). A complete list of NSF-funded curriculum projects can be found in National Science Foundation, *Course and Curriculum Improvement Materials: Mathematics, Science, Social Sciences* (Washington, DC: U.S. Government Printing Office, 1976).

10. On this particular point, "the myth of technical training," see Rudolph, *Scientists in the Classroom,* 167–171, and Rudolph, "Myth 23. That the Soviet Launch of *Sputnik* Caused the Revamping of American Science Education," 186–192.

11. Zacharias Interview, June 26, July 2, and July 16, 1975, PSSC Oral History Collection, Massachusetts Institute of Technology Archives; Judson Cross, "Notes on Teaching the New Physics Course," March 8, 1958, box 1, PSSC Oral History Collection; Bridgman, "The Prospect for Intelligence," 458. This view of the scientific method was shared by some of the biologists as well; see BSCS Steering Committee meeting minutes, May 13, 1961, Records of the Biological Science Curriculum Studies, Colorado Springs (hereafter BSCS Records).

12. Paulson to Wareham, October 24, 1955, NAS/CEP; Bates to Fuller, May 7, 1954, box 6, Marston Bates Papers, Bentley Historical Library, University of Michigan. On the changing nature of research in biology during this period, see Eric J. Vettel, *Biotech: The Countercultural Origins of an Industry* (Philadelphia: University of Pennsylvania Press, 2006); Toby A. Appel, *Shaping Biology: The National Science Foundation and American Biological Research, 1945–1975* (Baltimore: Johns Hopkins University Press, 2000); and Lily E. Kay, *The Molecular Vision of Life: Caltech, the Rockefeller Foundation, and the Rise of the New Biology* (New York: Oxford University Press, 1993).

13. Paul L. Dressel, "How the Individual Learns Science," in *Rethinking Science Education: The Fifty-Ninth Yearbook of the National Society for the Study of Education,* ed. Nelson B. Henry (Chicago: University of Chicago Press, 1960), 45–46.

14. Minutes of the first Steering Committee meeting, February 5, 1959, BSCS Records.

15. Quotations from "Fight for Mind's Freedom," *Science News Letter* 61 (January 12, 1952): 23, and Federation of American Scientists, Information Bulletin No. 38, April 17, 1954, 2, box 43, Records of the Federation of American Scientists, Department of Special Collections, University of Chicago. For a complete historical account of these attacks on scientists, see Jessica Wang, *American Science in an Age of Anxiety: Scientists, Anticommunism, and the Cold War* (Chapel Hill: University of North Carolina Press, 1999); Ellen W. Schrecker, *No Ivory Tower: McCarthyism and the Universities* (New York: Oxford University Press, 1986), 126–160; Lawrence Badash, "Science and McCarthyism," *Minerva* 38 (2000): 53–80; and David Kaiser, "The Atomic Secret in Red Hands? American Suspicions of Theoretical Physicists during the Early Cold War," *Representations* 90 (2005): 28–60.

16. Joel H. Hildebrand, "The Professor and His Public," *Bulletin of the Atomic Scientists* 9 (1953): 25; Karl T. Compton, "Science and Security," *Bulletin of the Atomic Scientists* 4 (1948): 375; "Free Speech in Science," *New York Times*, April 16, 1950, E8; "Battle of Ideas Must Be Won First, Caltech Head Asserts," *Los Angeles Times*, June 10, 1950, 2; "Free Inquiry Termed Shield of Democracy," *New York Times*, September 26, 1952; "Sees Gain by Public: Conant Expects Comprehension of Scientific Inquiry Soon," *New York Times*, April 23, 1952. The importance of the association of intellectual freedom with scientific practice during this period is wonderfully described in Audra J. Wolfe, *Freedom's Laboratory: The Cold War Struggle for the Soul of Science* (Baltimore: Johns Hopkins University Press, 2018).

17. Warren Weaver, "AAAS Policy," *Scientific Monthly* 73 (1951): 335–336; Edward Shils, "The Scientific Community: Thoughts after Hamburg," *Bulletin of the Atomic Scientists* 10 (1954): 151 (much of this work, it should be noted, was part of CIA-funded efforts to combat international communism through covert cultural diplomacy; see Wolfe, *Freedom's Laboratory*); Minutes of the APA Committee on the Freedom of Enquiry, September 5, 1954, box 35, Records of the Federation of American Scientists; Edwin G. Boring, "Report of the Committee on Freedom of Enquiry," *American Psychologist* 10 (1955): 289–293.

18. David A. Hollinger, "Free Enterprise and Free Inquiry: The Emergence of Laissez-Faire Communitarianism in the Ideology of Science in the United States," *New Literary History* 21 (1990): 897–919; Jamie Cohen-Cole, "The Creative American: Cold War Salons, Social Science, and the Cure for Modern Society," *Isis* 100 (2009): 219–262; Waldemar Kaempffert, "The Atom and the Scientific Mind," *New York Times*, October 9, 1949; Sidney Hook, "Academic Integrity and Academic Freedom," *Commentary* 8 (1949): 337; see also Aaron J. Ihde, "Responsibility of the Scientist to Society," *Scientific Monthly* 77 (1953): 244–249.

19. Bentley Glass, "Academic Freedom and Tenure in the Quest for National Security," *Bulletin of the Atomic Scientists* 12 (1956): 221.

20. On the dangers of military funding, see Bentley Glass, "The Academic Scientist, 1940–1960," *Science* 132 (1960): 598–603; and Glass, "Scientists in Politics," *Bulletin of the Atomic Scientists* 18 (1962): 2–7. Lysenkoism was something that was of particular concern to American biologists of the time; see Wolfe, *Freedom's Laboratory.* Quotation from H. Bentley Glass, "The Responsibilities of Biologists," *AIBS Bulletin* 7, no. 5 (1957): 13. In this talk, Glass reiterated the comments he had made in an earlier piece of work: Glass, "Science and Human Freedom," in *The Inauguration of Otto Frederick Kraushaar as President of Goucher College* (Towson, MD: Goucher College, 1949), 34–38. On Glass's public role during this time, see Audra Jayne Wolfe, "Speaking for Nature and Nation: Biologists as Public Intellectuals in Cold War Culture," PhD diss., University of Pennsylvania, 2002.

21. BSCS Steering Committee meeting minutes, February 5, 1959, and AIBS, "Proposal to the National Science Foundation for a Continuation of the Secondary School Program of the Biological Sciences Curriculum Study," April 29, 1960, both BSCS Records.

22. BSCS Steering Committee meeting minutes, January 28–29, 1960, and Moore to BSCS Steering Committee, memorandum, January 16, 1960, both BSCS Records.

23. BSCS Memorandum No. 32, November 24, 1959, box 53, Bentley Glass Papers, American Philosophical Society.

24. Donald N. Levine, *Powers of the Mind: The Reinvention of Liberal Learning in America* (Chicago: University of Chicago Press, 2006), 114–145; Ian Westbury and Neil J. Wilkof, eds., *Joseph J. Schwab: Science, Curriculum, and Liberal Education: Selected Essays* (Chicago: University of Chicago Press, 1978), 1–14.

25. Joseph J. Schwab, "Deriving the Objectives and Content of the College Curriculum: The Natural Sciences," in *New Frontiers in Collegiate Instruction,* ed. John Dale Russell (Chicago: University of Chicago Press, 1941), 46.

26. John W. Boyer, *A Twentieth-Century Cosmos: The New Plan and the Origins of General Education at Chicago,* Occasional Papers on Higher Education vol. 16 (Chicago: College of the University of Chicago, 2006), 139; Merle C. Coulter, Zens L. Smith, and Joseph J. Schwab, "The Science Programs in the College of the University of Chicago," in *Science in General Education,* ed. Earl J. McGrath (Dubuque, IA: Wm. C. Brown, 1948), 71.

27. Coulter, Smith, and Schwab, "The Science Programs in the College of the University of Chicago," 72–74.

28. Coulter, Smith, and Schwab, "The Science Programs in the College of the University of Chicago," 75 (emphasis in original).

29. John Dewey, *Logic: The Theory of Inquiry* (New York: Henry Holt, 1938), 8, 9; Joseph J. Schwab, "The Structure of the Disciplines: Meanings and Significances," in *The Structure of Knowledge and the Curriculum,* ed. G. W. Ford and Lawrence Pugno (Chicago: Rand McNally, 1964), 24. The influence of Dewey on Schwab's

work is described in the introduction to Westbury and Wilkof, eds., *Joseph J. Schwab: Science, Curriculum, and Liberal Education*, 1–40, esp. 36–39. Dewey, for his part, welcomed Schwab's work in this area. He wrote to a colleague that found it "in harmony" with his own work; John Dewey to Arthur F. Bentley, October 3, 1949 (document 15786), Dewey Correspondence.

30. Joseph J. Schwab, "The Structure of the Natural Sciences," in *The Structure of Knowledge and the Curriculum*, ed. G. W. Ford and Lawrence Pugno (Chicago: Rand McNally, 1964), 31. His efforts to get at the details of scientific work were also evident in Joseph J. Schwab, "What Do Scientists Do?," *Behavioral Science* 5 (1960): 1–27.

31. Schwab, "The Structure of the Natural Sciences," 31–49. Schwab's description of the process of science as proceeding through alternating periods of stable and fluid inquiry were, interestingly, nearly identical to Thomas Kuhn's description of normal and revolutionary science that he described in *The Structure of Scientific Revolutions*, which was published at about the same time as Schwab's work. I have yet to find any acknowledgment of either's work by the other in the historical record.

32. Joseph J. Schwab, "The Teaching of Science as Inquiry," *Bulletin of the Atomic Scientists* 14 (1958): 375; Bentley Glass, "Liberal Education in a Scientific Age," *Bulletin of the Atomic Scientists* 14 (1958): 349.

33. Joseph J. Schwab, "The Teaching of Science as Enquiry," in *The Teaching of Science*, ed. Joseph J. Schwab and Paul F. Brandwein (Cambridge, MA: Harvard University, 1962), 3–103; Albert A. Michelson, "Some of the Objects and Methods of Physical Science," *Quarterly Calendar* 3, no. 2 (August 1894): 15.

34. Schwab, "The Teaching of Science as Enquiry," 19, 9. On the changes in physics, see Barbara Lovett Cline, *Men Who Made a New Physics: Physicists and the Quantum Theory* (Chicago: University of Chicago Press, 1987), and Richard Staley, *Einstein's Generation: The Origins of the Relativity Revolution* (Chicago: University of Chicago Press, 2008).

35. Schwab, "The Teaching of Science as Enquiry," 24, 4–5.

36. Schwab, "The Teaching of Science as Enquiry," 37–48.

37. Schwab, "The Teaching of Science as Enquiry," 45–47.

38. Boyer, *A Twentieth-Century Cosmos*, 17. On the importance of discussion as a form of pedagogy for Schwab, see Levine, *Powers of the Mind*, 129–133.

39. Schwab, "The Teaching of Science as Enquiry," 71.

40. Record of Interview, Subject: Harvard University Conference on the Place of Science in General Education, July 9–10, 1949, n.d., box 161, CCNY Records; Coulter, Smith, and Schwab, "The Science Programs in the College of the University of Chicago," 70 (emphasis in original), 82–83; Schwab, "The Teaching of Science as Enquiry," 75; and Joseph J. Schwab, "Dialectical Means vs. Dogmatic Extremes in Relation to Liberal Education," *Harvard Educational Review* 21 (1951): 37–64 (emphasis in original).

41. Schwab, "The Teaching of Science as Inquiry," 377–379; Glass, "The Pervading Biological Themes," June 21, 1961 (emphasis in original), box 5, Glass Papers.

42. Biological Science Curriculum Study, *High School Science* [green version] (Chicago: Rand McNally, 1963); Biological Science Curriculum Study, *Biological Science: An Inquiry into Life* [yellow version] (New York: Harcourt, Brace & World, 1963); Biological Sciences Curriculum Study, *Biological Science: Molecules to Man* [blue version] (Boston: Houghton Mifflin, 1963); Schwab, "Teaching of Science as Inquiry," 377. For more detail on the curriculum materials produced by BSCS, see Rudolph, *Scientists in the Classroom,* 137–164.

43. Joseph J. Schwab, ed., *Biology Teachers' Handbook* (New York: Wiley, 1963), 47–48 (emphasis in original), 48, 59–60, 130–135; BSCS Steering Committee meeting minutes, May 26–27, 1962, BSCS Records.

8. SCIENTISTS IN THE CLASSROOM

1. Text from BSCS, *High School Biology* [green version] (Chicago: Rand McNally, 1963), The Laboratory, Introduction, attachment to memorandum No. 137, Grobman to BSCS Steering Committee, July 24, 1963, box 10, Bentley Glass Papers, American Philosophical Society; "A Progress Report: The First Six Years of the Physical Science Study Committee, The First Four Years of Educational Services Incorporated," n.d., unpublished typescript (emphasis in original), box 9, Jerrold Zacharias Papers, Institute Archives and Special Collections, Massachusetts Institute of Technology.

2. Barbara Barksdale Clowse, *Brainpower for the Cold War: The Sputnik Crisis and National Defense Education Act of 1958* (Westport, CT: Greenwood Press, 1981); Wayne Urban, *More than Science and Sputnik: The National Defense Education Act of 1958* (Tuscaloosa: University of Alabama Press, 2010). The $300 million figure is from Urban, *More than Science and Sputnik,* 173; U.S. Department of Health, Education, and Welfare, Office of Education, *Report on the National Defense Education Act, Fiscal Year Ending June 30, 1959* (Washington, DC: U.S. Government Printing Office, 1960), 7.

3. "Biological Sciences Curriculum Study Revolutionizes Lab Work," *Boulder Camera,* July 20, 1961; Earl and Robert Ubell, "New Way to Teach Physics," *New York Herald Tribune,* December 4, 1960, F6; Fred M. Hechinger, "Physics Teaching Enters Atom Age," *New York Times,* November 16, 1959, 1, 36.

4. Cynthia Placek, "The New Biology," *Chicago Tribune,* August 28, 1966, H44.

5. Walter C. Michels, "Committee on High School Teaching Materials," May 31, 1956, box 1, PSSC Oral History Collection, Massachusetts Institute of Technology Archives; Physical Science Study Committee, "General Report," March 25, 1957, 5, box 17, ESI-PSSC-EDC Papers, Institute Archives and Special Collections, MIT (hereafter PSSC Papers); Physical Science Study Committee, Preliminary Report,

March 12, 1957, 3, box 16, Jerrold Zacharias Papers, Institute Archives and Special Collections, Massachusetts Institute of Technology; Physical Science Study Committee, General Report, March 25, 1957, box 17, PSSC Papers; Physical Science Study Committee, General Report of the Physical Science Study Committee, March 25, 1957 (revised August 20, 1957), 2, box 17, Zacharias Papers; Elbert P. Little, "From These Beginnings . . . ," *Science Teacher* 24 (1957): 318.

6. Cornell University, Departments of Physics and Chemistry, "A Two-Year Course in Physical Science for High Schools," ca. 1957, box 65, Hans Bethe Papers, Division of Rare and Manuscript Collections, Cornell University; Jerrold R. Zacharias, "Educational Methods and Today's Science—Tomorrow's Promise," *Technology Review* 59 (1957): 501.

7. Physical Science Study Committee, *First Annual Report of the Physical Science Study Committee,* vol. 1, preliminary ed. (Cambridge, MA: Physical Science Study Committee, ca. 1957), 38; Alan T. Waterman Diary Note, March 10, 1959, box 25, Office of the Director Subject Files, Records of the National Science Foundation, National Archives and Records Administration, College Park, MD.

8. Jerrold R. Zacharias Oral Interview, June 26, July 2, and July 16, 1975, box 3, PSSC Oral History Collection; Physical Science Study Committee, *First Annual Report of the Physical Science Study Committee,* 8; Physical Science Study Committee (Uri Haber-Schaim), *Teacher's Guide for Laboratory Experiments* (Boston: D. C. Heath, 1960), 1; "A Progress Report: The First Six Years of the Physical Science Study Committee, The First Four Years of Educational Services Incorporated," n. d., 11, box 9, Zacharias Papers.

9. Physical Science Study Committee, *Laboratory Guide No. 1, Preliminary Edition* (Cambridge, MA: Physical Science Study Committee, 1958), 1, 2; J. B. Cross, The Physical Science Study Committee: Teaching the New Physics Course, presentation notes, March 8, 1958, box 1, PSSC Oral History Collection.

10. Physical Science Study Committee, *Laboratory Guide No. 1,* 2 (emphasis in original).

11. John Marean to Gilbert Finlay, memorandum, February 3, 1958, box 17, Zacharias Papers; PSSC Teachers Area Meeting, Drake University, Des Moines, IA, April 25, 1959, box 3, PSSC Papers. Quotation from Notes on Area Meeting Held at San Francisco State College, April 11, 1959; reference to students being lost in Educational Services Incorporated, Proposal to the National Science Foundation for the Support of the Continued Accumulation and Analysis of Feedback for the Period 1 January 1963–31 July 1963, January 31, 1963, box 5, PSSC Papers.

12. General Comments on Laboratory for Volume I, n.d.; Volume II—Latter Portion—Comments on Lab. in General, n.d., box 3, PSSC Papers.

13. General Comments on Laboratory for Volume I, n.d.; Volume II—Latter Portion—Comments on Lab. in General, n.d.; PSSC Report and Criticism, Bowdoin College, August 4, 1958 (emphasis in original), box 3; Notes on Area Meeting

Held at San Francisco State College, April 11, 1959, box 3; C. S. Randall to R. V. Bartz, F. L. Friedman, G. C. Finlay, E. P. Little, S. White, and J. R. Zacharias, "Partial Results of Teacher Survey Conducted by PSSC in May, 1958," memorandum, September 5, 1958, box 3, all PSSC Papers. John Marean to Gilbert Finlay, memorandum, February 3, 1958, box 17; "A Progress Report: The First Six Years of the Physical Science Study Committee, The First Four Years of Educational Services Incorporated," n.d., 14, box 9, Zacharias Papers.

14. University of California Berkeley to National Science Foundation, Organization of a Chemical Education Materials Study, 1 March 1960 to 30 September 1960, 3, box 471; J. Arthur Campbell, "A Possible Organization for a High School Chemistry Course," January 22, 1960, box 473; Chem Study Steering Committee Agenda, April 3, 1960, box 472, all Glenn T. Seaborg Papers, Library of Congress. The role of the laboratory in the Chemical Bond Approach Project was quite similar. See CBA Newsletter 9 (February 1961).

15. Chemical Education Material Study, Chemistry: An Experimental Science (San Francisco: W. H. Freeman, 1963), 1–8; Chemical Education Materials Study, Chemistry: An Experimental Science, Teachers Guide (San Francisco: W. H. Freeman, 1963), 2.

16. Richard J. Merrill and David W. Ridgway, The CHEM Study Story (San Francisco: W. H. Freeman, 1969), 29.

17. BSCS Steering Committee meeting minutes, June 9, 1959; BSCS Steering Committee meeting minutes, January 28, 1960, BSCS Records. For a summary of the block program, see Addison E. Lee, "The Block of Time Idea in Biology Laboratory Instruction," American Biology Teacher 22 (1960): 135–139.

18. Arnold B. Grobman to BSCS Steering Committee, memorandum no. 137, July 24, 1963, attachment 1, box 10.1, Glass Papers; Bentley Glass, "The Laboratory Block Program," typescript, 1959, BSCS Records; Committee on Innovation in Laboratory Instruction, meeting minutes, August 5–8, 1959, box 53.4, Glass Papers.

19. Descriptions of the traditional epistemology of natural history study are found in Keith R. Benson, "From Museum Research to Laboratory Research: The Transformation of Natural History into Academic Biology," in The American Development of Biology, ed. Ronald Rainger, Keith R. Benson, and Jane Maienschein (Philadelphia: University of Pennsylvania Press, 1988), 49–83, and David Magnus, "Down the Primrose Path: Competing Epistemologies in Early Twentieth-Century Biology," in Biology and Epistemology, ed. Richard Creath and Jane Maienschein (New York: Cambridge University Press, 2000), 91–121. Quotation from Committee on Innovation in Laboratory Instruction, meeting minutes, August 5–8, 1959, box 53.4, Glass Papers. Evelyn Fox Keller, "Physics and the Emergence of Molecular Biology: A History of Cognitive and Political Synergy," Journal of the History of Biology 23 (1990): 389–409; Lily E. Kay, "Problematizing Basic Research in Molecular Biology," in Private Science: Biotechnology and the Rise of the Molecular Sciences, ed. Arnold Thackray (Philadelphia: University of Pennsylvania Press, 1998), 20–38; Pnina G.

Abir-Am, "The Molecular Transformation of Twentieth-Century Biology," in *Science in the Twentieth Century,* ed. John Krige and Dominique Pestre (Amsterdam: Harwood Academic, 1997), 495–524. On the importance to members of the Steering Committee of conveying the experimental nature of biology, what they called "dynamic biology," over traditional, descriptive biology, see Charles Prosser to Arnold Grobman, excerpted in BSCS memo no. 28, October 26, 1959, box 9, Marston Bates Papers, Bentley Historical Library, University of Michigan; Wallace O. Fenn, "Front Seats for Biologists," typescript of speech given before the American Institute of Biological Sciences annual meeting, Stillwater, OK, August 29, 1960, box 30, Wallace O. Fenn Papers, Edward G. Miner Library, University of Rochester Medical Center, Rochester, NY; Joseph J. Schwab to BSCS Steering Committee, June 1, 1959, BSCS Records; and Secondary School Biological Sciences Film Series meeting minutes, October 8–9, 1958, box 2, Tracy M. Sonneborn Papers, Lilly Library, Indiana University.

20. Committee on Innovation in Laboratory Instruction, meeting minutes, August 5–8, 1959, box 53.4, Glass Papers.

21. Addison E. Lee, "The Use of BSCS Laboratory Blocks in a Modern Biology Program," box 43.2; "Discussing Laboratory Work in Biology," undated memorandum, box 6.1; Addison E. Lee, "The Use of BSCS Laboratory Blocks in a Modern Biology Program," box 43.2, Glass Papers.

22. Alfred S. Sussman, "Microbes: Their Growth, Nutrition and Interaction," *American Biology Teacher* 23 (1961): 411–418; BSCS Steering Committee meeting minutes, February 2, 1961, BSCS Records.

23. "Introduction to the Teacher's Guide Laboratory Exercises," undated draft, box 6.1 of Glass Papers; James T. Robinson, "Discussing Laboratory Work in Biology," undated typescript, box 6.1, Glass Papers.

24. BSCS Steering Committee meeting minutes, May 24, 1963, BSCS Records; Handwritten note by Don B. on Lab Project Summary "Project No. 164A: A Laboratory and Field Study of Seed Dispersal and Plant Distribution," box 5.2, Glass Papers; BSCS Steering Committee meeting minutes, May 24, 1963, BSCS Records.

25. Norman Abraham and Alfred Novak, "Observations on Laboratory Facilities for BSCS High School Biology," *BSCS Newsletter* 9 (1961): 8.

26. Committee on Innovation in Laboratory Instruction meeting minutes, June 12–13, 1959, box 50.3, Glass Papers; Richard E. Barthelemy, James R. Dawson, and Addison E. Lee, *Innovations in Equipment and Techniques for the Biology Teaching Laboratory* (Boston: D. C. Heath, 1964), 2; Paul Klinge, "The CCSSO Purchase Guide and BSCS Biology," and Margaret McKibben Lawler, "The BSCS and the National Defense Education Act (NDEA)," in *BSCS Biology—Implementation in the Schools,* BSCS Bulletin no. 3, ed. Hulda Grobman (Boulder, CO: Biological Sciences Curriculum Study, 1964), 62–65, 66–71; Barthelemy, Dawson, and Lee, *Innovations in Equipment and Techniques for the Biology Teaching Laboratory.*

27. Bert Kempers and Walter Auffenberg, "The Film Program," *BSCS Newsletter* 28 (1966): 10–11; Meeting Minutes, Learning Aids Committee, April 10, 1961, box 5.4, Glass Papers; "Suggested Program of Future Work of the BSCS Committee on Innovation in Laboratory Instruction," staff paper, August 1, 1961, box 5.4, Glass Papers. The list of training sessions is given in *BSCS Newsletter* 13 (1962): 5–6. Barthelemy, Dawson, and Lee, *Innovations in Equipment and Techniques for the Biology Teaching Laboratory;* quotation from BSCS Steering Committee meeting minutes, May 26, 1962, BSCS Records; Joseph J. Schwab, *Biology Teachers' Handbook* (New York: John Wiley and Sons, 1963).

28. Meeting Minutes, Committee on Innovation in Laboratory Instruction, June 2–4, 1961, box 43.2, Glass Papers.

29. BSCS Steering Committee meeting minutes, February 2, 1961, BSCS Records.

30. Abraham and Novak, "Observations on Laboratory Facilities for BSCS High School Biology," 8–12; BSCS Steering Committee meeting minutes, February 2, 1961, BSCS Records.

31. Committee on Innovation in Laboratory Instruction, meeting minutes, June 12–13, 1959, box 50.3, Glass Papers. On the history of Ward's, see Sally Gregory Kohlstedt, "Henry A. Ward: The Merchant Naturalist and American Museum Development," *Journal of the Society for the Bibliography of Natural History,* 1980, 647–661, and William C. Gamble, "Fossils to Plastics: A Century of Scientific Support," *AIBS Bulletin* 12, no. 3 (1962): 28–30. An overview of this focus on visual display of whole organisms is given in William C. Gamble, "The Role of the Biologist in Biological Supply Houses," *AIBS Bulletin* 13, no. 4 (1963): 38–40. The definitive history of display in science and natural history museums is found in Karen A. Rader and Victoria E. M. Cain, *Life on Display: Revolutionizing U.S. Museums of Science and Natural History Museums in the Twentieth Century* (Chicago: University of Chicago Press, 2014).

32. Information on previewing the BSCS materials is from a letter from William C. Gamble to Dean L. Gamble, December 12, 1962, box 23, Ward's Natural Science Establishment Papers, Department of Rare Books and Special Collections, University of Rochester (hereafter Ward's Papers); quotations are from William C. Gamble to Dean L. Gamble, April 18, 1962, box 23, and William C. Gamble to Dean L. Gamble, March 29, 1963, box 36, Ward's Papers.

33. John L. Rudolph, "Teaching Materials and the Fate of Dynamic Biology in American Classrooms after Sputnik," *Technology & Culture* 53 (2012): 1–36.

34. Rudolph, "Teaching Materials and the Fate of Dynamic Biology," 1–36.

35. Stanley L. Helgeson, Patricia E. Blosser, and Robert W. Howe, *The Status of Pre-College Science, Mathematics, and Social Science Education: 1955–1975,* vol. 1, *Science Education* (Washington, DC: National Science Foundation, 1977), 23–29; Wayne W. Welch, "Evaluating the Impact of National Curriculum Projects," *Science*

Education 60 (1976): 477. Frequency of "inquiry approach" from Google Books Ngram viewer.

36. Stephen G. Brush, "History of Science and Science Education," *Interchange* 20, no. 2 (1989): 61. It was Holton's book *Introduction to Concepts and Theories in Physical Science* (Cambridge, MA: Addison-Wesley, 1952), based on the Nat. Sci. 2 course, that inspired Rutherford to take up the history of science and consider further study at Harvard in the interest of science education reform; David Meshoulam, "'Teaching Physics as One of the Humanities': The History of (Harvard) Project Physics," PhD diss., University of Wisconsin–Madison, 2014. Interestingly, Kemble's own graduate work at Harvard in physics never would have occurred without a personal fellowship provided by Wallace Sabine, who was instrumental in producing the laboratory apparatus for Edwin Hall's physics course that was famously displayed at the World's Columbian Exposition in Chicago in 1893. Samuel H. Beer, Roy J. Glauber, Edward M. Purcell, and Gerald Holton, "Edwin C. Kemble" [obituary], *Physics Today* 40, no. 9 (1987): 97–99.

37. Linda Ingison, *The Status of Pre-College Science, Mathematics, and Social Studies Educational Practices in U.S. Schools: An Overview and Summaries of Three Studies* (Washington, DC: National Science Foundation, 1978), foreword.

38. Helgeson, Blosser, and Howe, *Status of Pre-College Science, Mathematics, and Social Science Education,* 38, 168, 175; Robert E. Stake and Jack A. Easley Jr., *Case Studies in Science Education,* vol. II, *Design, Overview and General Findings* (Washington, DC: National Science Foundation, 1978), cover letter to the project.

39. Margaret S. Steffensen, "Chapter 12: The Various Aims of Science Education," in *Case Studies in Science Education,* ed. Stake and Easley, 12:4 (emphasis in original).

40. Margaret S. Steffensen, "Chapter 12: The Various Aims of Science Education," in *Case Studies in Science Education,* ed. Stake and Easley, 12:4, 12:5.

41. Stake and Easley, eds., *Case Studies in Science Education,* 13:5, 13:14, 19:3, 19:14, 13:63.

42. Stake and Easley, eds., *Case Studies in Science Education,* 19:6 (emphasis in original).

43. Stake and Easley, eds., *Case Studies in Science Education,* 13:7; Steffensen, "Chapter 12: The Various Aims of Science Education," 12:7.

9. PROJECT 2061 AND THE NATURE OF SCIENCE

1. Diane Ravitch, *Left Back: A Century of Failed School Reforms* (New York: Simon & Schuster, 2000), 371–407; Kelly Moore, *Disrupting Science: Social Movements, American Scientists, and the Politics of the Military, 1945–1975* (Princeton, NJ: Princeton University Press, 2008), 1–16.

2. "Rutherford to Be AAAS Advisor," press release, January 4, 1981; "A Time for Change," *Science Education News,* November 1982, 1, Project 2061 Papers,

American Association for the Advancement of Science Archives, Washington, DC; quoted in "AAAS Charts Actions to Improve Science Literacy," *Science, Technology, and Human Values* 6, no. 35 (1981): 39.

3. Rutherford to David Robinson, September 21, 1982; "Question of Science Literacy Addressed by Rutherford / AAAS," April 1981, Project 2061 Papers.

4. F. James Rutherford, "Sputnik, Halley's Comet, and Science Education," unpublished draft manuscript, August 23, 1982, Project 2061 Papers. Lamenting the lack of educational initiatives at the time, Ernest Boyer, the commissioner of education under President Carter, quipped, "Maybe what we should do is get the Japanese to put a Toyota into orbit"; Edward B. Fiske, "Sputnik Recalled: Science and Math in Trouble Again," *New York Times,* October 5, 1982.

5. Quotes from "Project 2061: Phase I," proposal, October 15, 1985, Project 2061 Papers. Information on Halley's Comet from Rutherford, "Sputnik, Halley's Comet, and Science Education." The original name of the project was actually Project 2062, after a year that appeared in Gwyneth Cravens, "Toasting Halley's Comet," *New Yorker,* June 27, 1983, 76–79, but was changed to 2061 after consultation with astronomers who calculated the actual year of return to be 2061; see F. James Rutherford memorandum, Subject: Halley's Comet, December 21, 1984, Project 2061 Papers.

6. Rutherford to Alden Dunham, January 3, 1984, Project 2061 Papers; American Association for the Advancement of Science, *Science for All Americans: A Project 2061 Report on Literacy Goals in Science, Mathematics, and Technology* (Washington, DC: American Association for the Advancement of Science, 1989).

7. Daniel J. Kevles, *The Physicists: The History of a Scientific Community in Modern America* (Cambridge, MA: Harvard University Press, 1987), 393–409; Moore, *Disrupting Science;* Eric J. Vettel, *Biotech: The Countercultural Origins of an Industry* (Philadelphia: University of Pennsylvania Press, 2008); Allan Mazur, "Commentary: Opinion Poll Measurement of American Confidence in Science," *Science, Technology, and Human Values* 6, no. 36 (1981): 16–17; Boyce Rensberger, "The Invasion of the Pseudoscientists," *New York Times,* November 20, 1977, 16. Alarmed by the shifting public mood, the National Science Board (the oversight body of the National Science Foundation) initiated a comprehensive evaluation of the state of science in the United States, launching the first of an ongoing series of reports in 1972 that tracked what it called leading "science indicators." These included data on, among other things, federal support for research, the production of graduate degrees, and public attitudes toward science; National Science Board, *Science Indicators 1972* (Washington, DC: National Science Board, 1973).

8. Gene I. Maeroff, "A Report Finds Scientific Knowledge Has Declined among Pupils in U.S.," *New York Times,* June 22, 1974, 32; "Low Scores in Science Tests: Another Problem for the Schools," *U.S. News and World Report,* June 9, 1975, 43; Frederick V. Boyd and Elaine Sciolino, "The Decline in SAT Scores," *Newsweek,* March 8, 1976, 58; Ravitch, *Left Back,* 408–411.

9. William D. Hartley, "Changing Challenge: Japan Begins to Make More Complex Goods to Compete with West: Nation Aims to Manufacture Products Depending More on Brains than on Prices," *Wall Street Journal,* February 9, 1972, 1; James Reston, "Japan's Economic Invasion," *New York Times,* November 25, 1977, 25; James T. Patterson, *Restless Giant: The United States from Watergate to* Bush v. Gore (New York: Oxford University Press, 2005), 7, 62–65.

10. Thomas O'Toole, "U.S. Seen Losing Technological Edge in Some Industries," *Washington Post,* November 24, 1978, A14. Background on the rise of innovation as key to economic growth in the United States is found in Elizabeth Popp Berman, *Creating the Market University: How Academic Science Became an Economic Engine* (Princeton, NJ: Princeton University Press, 2012), 40–57.

11. Kevles, *The Physicists,* 412–413; Jerome Wiesner, "The Rise and Fall of the President's Science Advisory Committee," in *Jerry Wiesner: Scientist, Statesman, Humanist: Memories and Memoirs,* ed. Walter A. Rosenblith (Cambridge, MA: MIT Press, 2003), 408. Quotation from Daniel S. Greenberg, "Science and Richard Nixon," *New York Times Magazine,* June 17, 1973, 230.

12. Edward J. Larson, *Trial and Error: The American Controversy over Creation and Evolution,* 3rd ed. (New York: Oxford University Press, 2003), 125–155; Ronald L. Numbers, *The Creationists: From Scientific Creationism to Intelligent Design,* expanded ed. (Cambridge, MA: Harvard University Press, 2006), 268–279; Christopher J. Phillips, *The New Math: A Political History* (Chicago: University of Chicago Press, 2015), 122–143.

13. Erika Lorraine Milam, "Public Science of the Savage Mind: Contesting Cultural Anthropology in the Cold War Classroom," *Journal of the History of the Behavioral Sciences* 49 (2013): 306–330; Jamie Cohen-Cole, *The Open Mind: Cold War Politics and the Sciences of Human Nature* (Chicago: University of Chicago Press, 2014), 190–214. Details of the controversy over MACOS and the NSF education program as a whole can be found in Peter Dow, *Schoolhouse Politics: Lessons from the Sputnik Era* (Cambridge, MA: Harvard University Press, 1991), 178–238; see also Milam, "Public Science of the Savage Mind," 306–330.

14. Jimmy Carter, "The State of the Union: Annual Message to the Congress, January 19, 1978," *Public Papers of the Presidents of the United States: Jimmy Carter, 1978* (Washington, DC: U.S. Government Printing Office, 1979), 98–123.; Frank Press, Remarks at the 144th National Meeting of the American Association for the Advancement of Science, February 13, 1978, box 656, David A. Hamburg Papers, Rare Book and Manuscript Library, Columbia University (hereafter Hamburg Papers); National Science Foundation, National Center for Science and Engineering Statistics, *National Patterns of R&D Resources: 2011–12 Data Update,* NSF 14-304, 2013, http://www.nsf.gov/statistics/nsf14304 (accessed October 2, 2017).

15. F. James Rutherford, "The Birth of *Science for All Americans,*" Project 2061 website, http://www.project2061.org/publications/2061connections/2009/media

/SFAA%20Origins.pdf (accessed September 14, 2016); see also "Rutherford to Be AAAS Advisor," press release, January 4, 1981, Project 2061 Papers.

16. Rutherford, "Sputnik, Halley's Comet, and Science Education"; "Question of Science Literacy Addressed by Rutherford / AAAS," news memorandum, April 1981, Project 2061 Papers; David Meshoulam, "'Teaching Physics as One of the Humanities': The History of (Harvard) Project Physics, 1961–1970," PhD diss., University of Wisconsin–Madison, 2014, 68, 91; F. James Rutherford to David Robinson, August 2, 1982, and September 21, 1982, Project 2061 Papers.

17. "Carnegie Corporation Announces New Program Directions," press release, January 31, 1984, Project 2061 Papers. Quotations from Conference on Education and Economic Growth, Agenda Statement, February 2, 1983, and "Carnegie Forum on Education and the Economy: Background Statement," undated typescript, ca. 1983, Project 2061 Papers. Among the many reports were the following: National Science Foundation, *What Are the Needs in Precollege Science, Mathematics, and Social Science Education? Views from the Field* (Washington, DC: National Science Foundation, 1980); American Association for the Advancement of Science, *Education in the Sciences: A Developing Crisis* (Washington, DC: AAAS, 1982); U.S. Department of Education, National Commission on Excellence in Education, *A Nation at Risk: The Imperative for Educational Reform* (Washington, DC: U.S. Government Printing Office, 1983); Paul E. Peterson, *Making the Grade: Report of the Twentieth Century Fund Task Force on Federal Elementary and Secondary Education Policy* (New York: Twentieth Century Fund, 1983); Education Commission of the States, Task Force on Education for Economic Growth, *Action for Excellence: A Comprehensive Plan to Improve Our Nation's Schools* (Denver, CO: Education Commission of the States, 1983); College Entrance Examination Board, *Academic Preparation for College: What Students Need to Know and Be Able to Do* (New York: College Board Office of Academic Affairs, 1983); National Science Board, *Educating Americans for the 21st Century* (Washington, DC: National Science Foundation, 1983).

18. U.S. Department of Education, National Commission on Excellence in Education, *A Nation at Risk,* 5; for recommendations, see 24–33. See also Maris A. Vinovskis, *From* A Nation at Risk *to* No Child Left Behind: *National Education Goals and the Creation of Federal Education Policy* (New York: Teachers College Press, 2009), 14–15; David A. Hamburg to Participants in June 29 Meeting, memorandum, June 20, 1983, Project 2061 Papers.

19. "Carnegie Forum on Education and the Economy: Background Statement," typescript, ca. 1983, Project 2061 Papers; "Excerpts from Education in the Sciences: A Developing Crisis," *Science Education News,* November 1982, 5; *Daedalus* 112, no. 2 (1983). Although one of the essays in that issue explored the notion from an economic perspective, the rest emphasized science education for humanistic and civic goals. The economic essay was Herbert J. Walberg, "Scientific Literacy and Economic Productivity in International Perspective," *Daedalus* 112, no 2 (1983): 1–28. On the history

of the phrase, see George E. DeBoer, "Scientific Literacy: Another Look at Its Historical and Contemporary Meanings and Its Relationship to Science Education Reform," *Journal of Research in Science Teaching* 37 (2000): 582–601.

20. "Project 2061: Understanding Science and Technology for Living in a Changing World," proposal by the American Association for the Advancement of Science, March 30, 1985, 18–19, Project 2061 Papers; "Question of Science Literacy Addressed by Rutherford/AAAS," news memorandum, April 1981, Project 2061 Papers; "Project 2061: Understanding Science and Technology for Living in a Changing World," proposal by the American Association for the Advancement of Science, March 30, 1985, 18, Project 2061 Papers; "Excerpts from Education in the Sciences," 5.

21. F. James Rutherford to E. Alden Dunham, September 6, 1984, Project 2061 Papers.

22. "Excerpts from Education in the Sciences," 5 (emphasis in original); Grant Recommendation, April 12, 1985, box 1063A, CCNY Records; Sara L. Engelhardt to William Carey, June 21, 1985, box 9, Project 2061 Papers; Technology Panel notes for meeting held in Washington, DC, submitted by Karen M. Olson, January 13, 1986, Project 2061 Papers.

23. "Project 2061: Understanding Science and Technology for Living in a Changing World," proposal by the American Association for the Advancement of Science, March 30, 1985, 18 (emphasis in original), Project 2061 Papers.

24. "Project 2062: What Science Is Most Worth Knowing," draft prospectus, April 16, 1984, Project 2061 Papers; "Conference to Examine Plans for Determining the 'Fundamentals' of Science Education for All High School Graduates," Ref. JR letter to Alden Dunham, January 3, 1984, Project 2061 Papers; AAAS, *Science for All Americans,* 19.

25. AAAS Proposal for Project 2061, grant memorandum, December 5, 1984, box 1063A, CCNY Records; Bill Carey to F. James Rutherford, January 16, 1985, Project 2061 Papers.

26. "Project 2061: Understanding Science and Technology for Living in a Changing World," proposal by the American Association for the Advancement of Science, March 30, 1985, 43, Project 2061 Papers; David Hawkins to F. James Rutherford, February 8, 1985, box 9, Project 2061 Papers. The physical science panel was the one exception. It included an extended discussion of scientific process but agreed later on that the material it drafted would be better placed in an introductory chapter that would set the stage for all the content areas.

27. Quotations from F. James Rutherford, "The Role of Inquiry in Science Teaching," *Journal of Research in Science Teaching* 2 (1964): 80–84; Rutherford to Alden Dunham, January 3, 1984, Project 2061 Papers.

28. Physical Science and Engineering Panel, meeting minutes, January 20, 1986; National Council on Science and Technology Education, meeting transcript, January

13, 1986, 19–20; G. Bugliarello, Report of the Panel on Physical Sciences and Engineering, January 7, 1986; prepared notes for the Physical Science and Engineering Panel, minutes, December 11, 1985; Panel on Physical Science and Engineering, meeting minutes, November 14, 1985, all Project 2061 Papers.

29. Mary E. Clark, "Some Basic Ideas from 2061," National Academy of Sciences, January 13–14, 1986; National Council on Science and Technology Education, meeting transcript, January 14, 1986, 22; workshop sessions, meeting transcript, January 14, 1986, 263–264, all Project 2061 Papers.

30. AAAS, *Science for All Americans,* 26.

31. AAAS, *Science for All Americans,* 27–28.

32. F. James Rutherford, "Chapter Two: General Perspectives on Science, Mathematics, and Technology," Draft 4, September 2, 1986, Project 2061 Papers; AAAS, *Science for All Americans,* 28–31.

33. The tremendous growth of science as an enterprise at this time was highlighted by Derek J. de Solla Price's *Little Science, Big Science* (New York: Columbia University Press, 1963), which was itself an example of the new studies of science taking place. On the history of science studies as a field, see David Edge, "Reinventing the Wheel," in *Handbook of Science and Technology Studies,* rev. ed., ed. Sheila Jasanoff, Gerald E. Markle, James C. Peterson, and Trevor Pinch (Thousand Oaks, CA: Sage, 1995), 3–24; Steve Fuller, *Thomas Kuhn: A Philosophical History for Our Times* (Chicago: University of Chicago, 2000), 3, 318–331; and Lorraine Daston, "Science Studies and the History of Science," *Critical Inquiry* 35 (2009): 798–813. On the early history-of-science departments, see Victor L. Hilts, "History of Science at the University of Wisconsin," *Isis* 75 (1984): 63–94, and I. Bernard Cohen, "The *Isis* Crisis and the Coming of Age of the History of Science Society, with Notes on the Early Days of the Harvard Program in History of Science," *Isis* 90, supp. (1999): S28–S42.

34. Christopher Hamlin, "The Pedagogical Roots of the History of Science: Revisiting the Vision of James Bryant Conant," *Isis* 107 (2016): 284; F. James Rutherford, file memorandum, March 28, 1984, Project 2061 Papers. On Rutherford's background, see Meshoulam, "'Teaching Physics as One of the Humanities,'" 57–67.

35. Quotation from Spencer R. Weart to F. James Rutherford, September 17, 1986, Project 2061 Papers. Weart, a historian of science, served on the physical science panel, and various drafts were circulated for feedback among scholars such as philosophers of science Patrick Suppes and Philip Kitcher as well as historians of science Stephen Brush, Edwin Layton, Owen Gingerich, and Sally Gregory Kohlstedt, among others. On the various references to Kuhn and other science-studies scholars, see Mary Clark to Project 2061 Biology Panel, November 8, 1986; G. Bugliarello, Notes from the Panel on the Physical Sciences and Engineering, January 7, 1986; Mary E. Clark to Pat Warren, Jim Rutherford, and Chick Ahlgren, July 16, 1986; Panel on Physical Science and Engineering Minutes, February 24, 1986, all Project 2061

Papers. Historian of technology Edwin Layton commented that most of the science sections seemed "old" but "generally good"; Layton to F. James Rutherford, January 6, 1987, Project 2061 Papers.

36. Each case provides a textbook example of what the sociologist Thomas Gieryn refers to as boundary work, that is, the use of definitions of science, often tied to some particular ideas about methodology, to include or exclude various intellectual or cultural practices from the category of science so that those practices might either benefit from or be denied the epistemic and cultural authority science enjoys among the broader public; Thomas F. Gieryn, *Cultural Boundaries of Science: Credibility on the Line* (Chicago: University of Chicago Press, 1999).

37. Thomas F. Gieryn, "The U.S. Congress Demarcates Natural Science and Social Science (Twice)," in *Cultural Boundaries of Science,* 65–114. See also Andrew Jewett, *Science, Democracy, and the American University: From the Civil War to the Cold War* (New York: Cambridge University Press, 2012), 310–320; and Mark Solovey, "Riding Natural Scientists' Coattails onto the Endless Frontier: The SSRC and the Quest for Scientific Legitimacy," *Journal of the History of the Behavioral Sciences* 40 (2004): 393–422.

38. David A. Hamburg, "Statement by David A. Hamburg, Chairman, American Association for the Advancement of Science on the Fiscal Year 1987 National Science Foundation Budget Authorization Request," typescript, February 25, 1986, 20, box 657, Hamburg Papers. Quotation from "Project 2061: Understanding Science and Technology for Living in a Changing World," proposal by the American Association for the Advancement of Science, March 30, 1985, 5, Project 2061 Papers.

39. "Basic Concepts in the Social and Behavioral Science: A Report of the Panel on Social and Behavioral Sciences. Project 2061: Education for a Changing Future," unpublished report, box 1061, CCNY Records.

40. "Basic Concepts in the Social and Behavioral Science: A Report of the Panel on Social and Behavioral Sciences."

41. National Council on Science and Technology Education, meeting transcript, June 4–5, 1987, 201–239, Project 2061 Papers.

42. AAAS, *Science for All Americans,* 29, 77.

43. Larson, *Trial and Error,* 147–155.

44. Larson, *Trial and Error,* 134–139; Kenneth M. Pierce, "Putting Darwin back in the Dock," *Time,* March 16, 1981, 80–82; "Reagan Proposes School Prayer Amendment," *New York Times,* May 18, 1982, 24; John Walsh, "At AAAS Meeting, a Closing of Ranks," *Science* 215 (January 22, 1982): 380; "AAAS: Forced Teaching of Creationist Beliefs in Public School Science Education," Adopted by the AAAS Board of Directors, January 4, 1982, and by the AAAS Council on January 7, 1982, http://archives.aaas.org/docs/resolutions.php?doc_id=361 (accessed October 27, 2016).

45. Agenda for Meeting of May 24, 1986, Salk Institute, Biology and Medicine Panel; National Council on Science and Technology Education, meeting transcript,

January 13, 1986, 165, and January 14, 1986, 121; Panel on the Physical Sciences and Engineering, "Report of the Physics and Mathematics Subpanel," typescript, April 25, 1986, all Project 2061 Papers.

46. F. James Rutherford, "General Perspectives on Science, Mathematics and Technology, the Framework," draft 4, September 2, 1986, 11, Project 2061 Papers.

47. Steven V. Roberts, "White House Confirms Reagans Follow Astrology, up to a Point," *New York Times*, May 4, 1988, A1; National Council on Science and Technology Education, meeting transcript, June 4–5, 1987, 69, 72, Project 2061 Papers.

48. AAAS, *Science for All Americans*, 25–26.

49. National Council on Science and Technology Education, meeting transcript, June 4–5, 1987, 67–68, Project 2061 Papers; AAAS, *Science for All Americans*, 118.

50. Chester E. Finn Jr., "The Science of Bad Science," *American Spectator* 22, no. 8 (August 1989): 34 (emphasis in original); National Council on Science and Technology Education, meeting transcript, January 13, 1986, 49, Project 2061 Papers; J. Myron Atkin, "Summary of Main Points in Ten-Minute Oral Testimony," Hearing on the FY1989 National Science Foundation Authorization, March 22, 1988, box 1063A, CCNY Records.

10. SCIENCE IN THE STANDARDS ERA

1. Thomas L. Friedman, *The World Is Flat: A Brief History of the Twenty-First Century* (New York: Farrar, Straus and Giroux, 2005), 10.

2. Quotations from Warren Bass, "The Great Leveling," review of *The World Is Flat: A Brief History of the Twenty-First Century* by Thomas L. Friedman, *Washington Post*, April 3, 2005, and *How the Lack of Higher Education Faculty Contributes to America's Nursing Shortage: Field Hearing before the Select Committee on Education*, House of Representatives, 109th Cong., November 30, 2005 (statement of Kay Norton); other references to Friedman's book found in, among many others, *Challenges to American Competitiveness in Math and Science: Hearing before the Committee on Education and the Workforce*, House of Representatives, 109th Cong. (May 19, 2005); *Science, Technology and Global Economic Competitiveness: Hearing before the Committee on Science*, House of Representatives, 109th Cong., October 20, 2005 (statement of Norman R. Augustine, retired chairman and chief executive officer, Lockheed Martin Corporation); *Sharpening Our Edge: Staying Competitive in the 21st Century Marketplace: Hearing before the Committee on Government Reform*, House of Representatives, 109th Cong. (February 9, 2006); and *K–12 Science and Math Education across the Federal Agencies: Hearing before the Committee on Science*, House of Representatives, 109th Cong., March 30, 2006.

3. Committee on Prospering in the Global Economy of the 21st Century, *Rising above the Gathering Storm: Energizing and Employing America for a Brighter Economic*

Future (Washington, DC: National Academies Press, 2007). The report was practically framed as a response to Friedman's book. Quotations from 1–3.

4. "Project 2061: Understanding Science and Technology for Living in a Changing World," proposal by the American Association for the Advancement of Science, March 30, 1985, 29 (emphasis in original), Project 2061 Papers; "Project 2061: Draft Plan for Phase II," draft, January 7, 1987, Project 2061 Papers; F. James Rutherford to Alden Dunham, February 17, 1988, box 1063A, CCNY Records. The district working teams (from rural Georgia; suburban McFarland, Wisconsin; urban Philadelphia, San Antonio, San Diego, and San Francisco) were charged as well with developing model curricula in their districts that would accomplish the learning goals laid out in the *Benchmarks*. These models failed to gain much traction as useful products of Project 2061.

5. "Chapter One: Learning and Teaching," draft, September 3, 1986, 1; JR, "Chapter One," very rough draft, May 13, 1986, 1; "Chapter One: Learning and Teaching," draft, September 3, 1986, 3–6, all Project 2061 Papers.

6. F. James Rutherford and Andrew Ahlgren, *Science for All Americans: A Project 2061 Report on Literacy Goals in Science, Mathematics, and Technology* (Washington, DC: American Association for the Advancement of Science, 1989), 145–151. Concerns over the marginalization of process were expressed in Spencer R. Weart to George Bugliarello and F. James Rutherford, October 2, 1986, and A. J. Harrison to George Bugliarello, March 19, 1987, both Project 2061 Papers.

7. Compiled reviewers' comments, folder "Chapter 0," 51, 12, 78, Project 2061 Papers.

8. Chick to FJR, "Are Benchmarks What We Have?," memorandum, March 2, 1993, Project 2061 Papers.

9. F. James Rutherford, "Hands-On," internal memorandum, March 4, 1993, Project 2061 Papers.

10. Rutherford, "Hands-On" (emphasis in original).

11. Compiled reviewers' comments, n.d., folder "Chapter 0," 80, 36, and F. James Rutherford to AA [Andrew Ahlgren], JER [Jo Ellen Roseman], and MW, "Response to AA 'Expectations for Benchmarks,'" memorandum, March 5, 1993, both Project 2061 Papers.

12. Project 2061, American Association for the Advancement of Science, *Benchmarks for Science Literacy* (New York: Oxford University Press, 1993), 312–313, 319–320, 9. Rutherford laid out the approach they took to the problems reviewers raised in "Response to AA 'Expectations for Benchmarks.'"

13. Project 2061, American Association for the Advancement of Science, *Benchmarks for Science Literacy*, 3–20; scientific method quotation is from 9.

14. Diane Ravitch, "The Search for Order and the Rejection of Conformity: Standards in American Education," in *Learning from the Past: What History Teaches Us about School Reform*, ed. Diane Ravitch and Maris A. Vinovskis (Baltimore, MD:

Johns Hopkins University Press, 1995), 180–181; Maris A. Vinovskis, *The Road to Charlottesville: The 1989 Education Summit* (Alexandria, VA: National Education Goals Panel, 1999), 40; Maris A. Vinovskis, *From* A Nation at Risk *to* No Child Left Behind: *National Education Goals and the Creation of Federal Education Policy* (New York: Teachers College Press, 2009), 27; Diane Ravitch, *Left Back: A Century of Failed School Reform* (New York: Simon & Schuster, 2000), 432; National Research Council, *National Science Education Standards* (Washington, DC: National Academy Press, 1996). Details on the chronology of events leading to the final document are given in Angelo Collins, "National Science Education Standards in the United States: A Process and a Product," *Studies in Science Education* 26 (1995): 7–37.

15. Bill G. Aldridge, "Project on Scope, Sequence, and Coordination: A New Synthesis for Improving Science Education," *Journal of Science Education and Technology* 1 (1992): 13–21; Constance Holden, "Demolishing the Layer Cake," *Science* 250 (1990): 205; J.T., "Editor's Corner: Scope, Sequence, and Coordination," *Science Teacher* 58 (1991): 6. The main tenets of the NSTA project can be found in the appendix of Bill G. Aldridge, ed., *Scope, Sequence, and Coordination: A Framework for High School Science Education* (Arlington, VA: National Science Teachers Association, 1996), 187. Angelo Collins, "National Science Education Standards: A Political Document," *Journal of Research in Science Teaching* 35 (1998): 715.

16. Andrew A. Zucker, Viki M. Young, and John M. Luczak, *Evaluation of the American Association for the Advancement of Science's Project 2061, Executive Summary* (Menlo Park, CA: SRI International, 1996); National Research Council, *National Science Education Standards,* 201, 200.

17. William J. Broad, "Big Science: Is It Worth the Price?," *New York Times,* May 27, 1990, 20. See also Malcolm W. Brown, "Downside to Ending of Cold War: Opportunities in Science Dwindle," *New York Times,* February 20, 1994, 36.

18. Paul R. Gross and Norman Levitt, *Higher Superstition: The Academic Left and Its Quarrels with Science* (Baltimore, MD: Johns Hopkins University Press, 1994). Kuhn had argued in *The Structure of Scientific Revolutions* that paradigm shifts in science—the revolutions that occur between periods of normal science—couldn't be accounted for solely by appeal to empirical evidence or even rationality. Accounts of the science wars include Dorothy Nelkin, "The Science Wars: Responses to a Failed Marriage," *Social Text* 46–47 (1996): 93–100; and Thomas F. Gieryn, *Cultural Boundaries of Science: Credibility on the Line* (Chicago: University of Chicago Press, 1999), 336–362. In the spring of the following year, Gross and Levitt (this time with the help of Gerald Holton) organized a conference at the New York Academy of Sciences titled *The Flight from Science and Reason,* the proceedings of which—extending the attack on the academic critics of science—were subsequently published in 1996; Paul R. Gross, Norman Levitt, and Martin W. Lewis, eds., *The Flight from Science and Reason* (New York: New York Academy of Sciences, 1996). Reference to *Higher Superstition* found in National Research Council, *National Science Education Standards,* 25.

19. Project 2061, American Association for the Advancement of Science, *Benchmarks for Science Literacy,* 312–313.

20. National Research Council, *National Science Education Standards,* 28–29.

21. National Research Council, *National Science Education Standards,* 105, 174–176.

22. On the rise of cognitive science, see Jamie Cohen-Cole, *The Open Mind: Cold War Politics and the Sciences of Human Nature* (Chicago: University of Chicago Press, 2014), 141–164. The impact of this work along with insights from science studies in science education is described in Richard Duschl, "Science Education in Three-Part Harmony: Balancing Conceptual, Epistemic, and Social Learning Goals," *Review of Research in Education* 32 (2008): 268–291. See also National Research Council, *How People Learn: Bridging Research and Practice* (Washington, DC: National Academy Press, 1999), and *Taking Science to School: Learning and Teaching Science in Grades K–8* (Washington, DC: National Academies Press, 2007), a report that extended the insights from *How People Learn* to science education.

23. National Research Council, *Inquiry and the National Science Education Standards: A Guide for Teaching and Learning* (Washington, DC: National Academy Press, 2000), 14–17, 34.

24. Gretchen Vogel, "The Calculus of School Reform," *Science* 277 (1997): 1192–1195; Angelo Collins, "National Science Education Standards: Looking Backward and Forward," *Elementary School Journal* 97 (1997): 309–310.

25. Zucker, Young, and Luczak, *Evaluation of the American Association for the Advancement of Science's Project 2061, Executive Summary,* viii; Karen S. Hollweg and David Hill, eds., *What Is the Influence of the National Science Education Standards? Reviewing the Evidence, a Workshop Summary* (Washington, DC: National Academies Press, 2003), 107; Joan D. Pasley, Iris R. Weiss, Elizabeth S. Shimkus, and P. Sean Smith, "Looking inside the Classroom: Science Teaching in the United States," *Science Educator* 13 (2004): 6; Iris R. Weiss, Joan D. Pasley, P. Sean Smith, Eric R. Banilower, and Daniel J. Heck, *Looking Inside the Classroom: A Study of K–12 Mathematics and Science Education in the United States* (Chapel Hill, NC: Horizon Research, 2003), H-23.

26. Zucker, Young, and Luczak, *Evaluation of the American Association for the Advancement of Science's Project 2061, Executive Summary,* 71; M. Patricia Morse and the AIBS Review Team, *A Review of Biological Instructional Materials for Secondary Schools* (Washington, DC: American Institute of Biological Sciences, 2001), 11, 1; Zucker, Young, and Luczak, *Evaluation of the American Association for the Advancement of Science's Project 2061, Executive Summary,* 18; Greta Vogel, "The Special Needs of Science," *Science* 277 (1997): 1193.

27. Horizon Research, "The Influence of the *National Science Education Standards* on Teachers and Teaching Practice," in *What Is the Influence of the National Science Education Standards? Reviewing the Evidence, a Workshop Summary,* ed.

Karen S. Hollweg and David Hill (Washington, DC: National Academies Press, 2003), 103–107; Hollweg and Hill, eds., *What Is the Influence of the National Science Education Standards?*, 9; Mark Windschitl, "Folk Theories of 'Inquiry': How Preservice Teachers Reproduce the Discourse and Practices of an Atheoretical Scientific Method," *Journal of Research in Science Teaching* 41 (2004): 481–512. For repeated references to the scientific method, see Ray T. Sterner, "The Scientific Method: An Instructor's Flow Chart," *American Biology Teacher* 60 (1998): 374–378; R. E. Uthe, "Projecting the Scientific Method," *Science Teacher* 67, no. 9 (2000): 44–47; and Annette Parrott, "Activities with Antlions: Students Learn Concepts of Insect Behavior by Observing Doodlebugs," *Science Teacher* 67 (2000): 51–53. On the presence of the traditional scientific method in textbooks, see James Blachowicz, "How Science Textbooks Treat Scientific Method: A Philosopher's Perspective," *British Journal for the Philosophy of Science* 60 (2009): 303–344.

28. Quotation from Hollweg and Hill, eds., *What Is the Influence of the National Science Education Standards?*, 6. The committee even devoted a fair amount of their report to that very history, examining the changing role laboratory work has played over the years beginning in the late 1800s. National Research Council, *America's Lab Report: Investigations in High School Science* (Washington, DC: National Academies Press, 2006), 9.

29. Education consultant quoted in Gretchen Vogel, "The Calculus of School Reform," 1194; "fact, fact, fact" quotation from *Improving the Laboratory Experience for America's High School Students, Hearing before the Subcommittee on Research and Science Education, Committee on Science and Technology,* House of Representatives, 110th Cong., March 8, 2007, 48. A comprehensive survey of state assessments found that they were poorly aligned with state standards, which routinely mirrored the NSES and *Benchmarks* emphasis on inquiry and the nature of science. This was particularly true in the area of "students' abilities to do investigations and achieve an understanding of the nature of science," which few tests were designed to measure. Norman L. Webb and Sarah A. Mason, "Taking Stock of the National Science Education Standards: The Research for Assessment and Accountability," in *What Is the Influence of the National Science Education Standards? Reviewing the Evidence, a Workshop Summary,* ed. Karen S. Hollweg and David Hill (Washington, DC: National Academies Press, 2003), 87. For a broad examination of the impact of testing on American education, see Diane Ravitch, *The Death and Life of the Great American School System: How Testing and Choice Are Undermining Education,* rev. ed. (New York: Basic Books, 2010).

30. Hollweg and Hill, eds., *What Is the Influence of the National Science Education Standards?*, 9; Paul A. Kirschner, John Sweller, and Richard E. Clark, "Why Minimal Guidance during Instruction Does Not Work: An Analysis of the Failure of Constructivist, Discovery, Problem-Based, Experiential, and Inquiry-Based Teaching," *Educational Psychologist* 41 (2006): 84, 83.

31. National Science Board, *Science and Engineering Indicators 2004* (Arlington, VA: National Science Foundation, 2004), ch. 1.

32. National Science Board, *Science and Engineering Indicators 2004*, 7–3; National Science Board, *An Emerging and Critical Problem of the Science and Engineering Labor Force: A Companion to Science and Engineering Indicators 2004* (Arlington, VA: National Science Foundation, 2004).

33. Tania Simoncelli with Jay Stanley, *Science under Siege: The Bush Administration's Assault on Academic Freedom and Scientific Inquiry* (New York: American Civil Liberties Union, 2005), foreword.

34. National Science Board, *An Emerging and Critical Problem of the Science and Engineering Labor Force;* Kenneth R. Foster and Irving A. Lerch, "Collateral Damage: American Science and the War on Terrorism," *IEEE Technology and Society Magazine* 24, no. 3 (2005): 45–53; see also Charles M. Vest, "Openness and Security: Problems, Progress, and Opportunity," speech, Annual Council of Presidents Luncheon of the National Association of State Universities and Land-Grant Colleges, New Orleans, November 17, 2003, http://web.mit.edu/president/communications/nasulgc03.html.

35. Friedman, *The World Is Flat,* quoted in William A. Wulf, "A Disturbing Mosaic," *The Bridge* 35, no. 3 (Fall 2005): 29; Committee on Prospering in the Global Economy of the 21st Century, *Rising above the Gathering Storm,* 2–7; William J. Broad, "Top Advisory Panel Warns of an Erosion of the U.S. Competitive Edge in Science," *New York Times,* October 13, 2005; J. Donald Capra, "Commentary: American Science Is in Hot Water," *Journal Record,* November 9, 2005; and "No Sputnik This Time, but a Crisis Just as Great; Congress Must Act Urgently to Preserve America's Edge; Advanced Energy Research Is One Good Place to Start," *San Jose Mercury News,* October 17, 2005.

36. George W. Bush, "Address before a Joint Session of the Congress on the State of the Union," January 31, 2006, The American Presidency Project, http://www.presidency.ucsb.edu/ws/index.php?pid=65090; *Building America's Competitiveness: Examining What Is Needed to Compete in a Global Economy, Hearing before the Committee on Education and the Workforce,* House of Representatives, 109th Cong., April 6, 2006, 15, 16; *Challenges to American Competitiveness in Math and Science,* 1 (concerns about visa restrictions were voiced by James Jarrett, a vice president at Intel, 46–48); *K–12 Science and Math Education across the Federal Agencies,* 10.

37. Committee on Prospering in the Global Economy of the 21st Century, *Rising above the Gathering Storm,* 5; American COMPETES Act, Public Law 110-69, 121 Stat. 572 (2007), sec. 6121.

38. Heather B. Gonzalez, *America COMPETES Acts: FY2008 to FY2013, Funding Tables,* CRS Report No. R42779 (Washington, DC: Congressional Research Service, 2014), https://fas.org/sgp/crs/misc/R42779.pdf; Deborah D. Stine, *America COMPETES Act: Programs, Funding, and Selected Issues,* CRS Report No. RL34328 (Washington, DC: Congressional Research Service, 2009), 26, http://www.stemedcoalition

.org/wp-content/uploads/2010/05/ADA501311-3.pdf (emphasis added); National Research Council, *Learning and Understanding: Improving Advanced Study of Mathematics and Science in U.S. High Schools* (Washington, DC: National Academy Press, 2002); William B. Wood, "Revising the AP Biology Curriculum," *Science* 325 (2009): 1627–1628; Serena Magrogan, "Past, Present, and Future of AP Chemistry: A Brief History of Course and Exam Alignment Efforts," *Journal of Chemical Education* 91 (2014): 1357–1361.

CONCLUSION

1. National Research Council, *A Framework for K–12 Science Education: Practices, Crosscutting Concepts, and Core Ideas* (Washington, DC: National Academies Press, 2012), ix. See also NGSS Lead States, *Next Generation Science Standards: For States, by States* (Washington, DC: National Academies Press, 2013).

2. NGSS website, https://www.nextgenscience.org/need-standards, accessed June 26, 2018; Carnegie Corporation of New York and Institute for Advanced Study Commission on Mathematics and Science Education, *The Opportunity Equation: Mobilizing for Excellence and Equity in Mathematics and Science Education for Citizenship and the Global Economy* (New York: Carnegie Corporation of New York, 2009).

3. Work in this area includes Andrew Pickering, *The Mangle of Practice: Time, Agency, and Science* (Chicago: University of Chicago Press, 1995); Jed Z. Buchwald, ed., *Scientific Practice: Theories and Stories of Doing Physics* (Chicago: University of Chicago Press, 1995); and Joseph Rouse, *Engaging Science: How to Understand Its Practices Philosophically* (Ithaca, NY: Cornell University Press, 1996). Cross-cutting concepts include things such as cause and effect, patterns, and scale, and disciplinary core ideas essentially are the main content of the various science disciplines. National Research Council, *A Framework for K–12 Science Education*, 3, 30.

4. National Research Council, *A Framework for K–12 Science Education*, 44.

5. It is typically the case that events that appear to be cycles of a pendulum swing are revealed on more careful historical examination to actually be entirely new events that have arisen within their own unique historical context. On the pendulum metaphor in education, see Herbert M. Kliebard, "Success and Failure in Educational Reform: Are There Historical 'Lessons'?," *Peabody Journal of Education* 65 (1988): 144–157.

6. Charles W. Eliot, "What Is a Liberal Education?," *Century Magazine* 28 (June 1884): 209.

7. Zacharias as quoted in Paul Forman, "Behind Quantum Electronics: National Security as Basis for Physical Research in the United States, 1940–1960," *Historical Studies in the Physical and Biological Sciences* 18 (1987): 152. Teller quoted by J. Myron Atkin, "Summary of Main Points in Ten-Minute Oral Testimony," Hearing on the FY1989 National Science Foundation Authorization, March 22, 1988, CC9.

8. Michael S. Teitelbaum, *Falling Behind? Boom, Bust, and the Global Race for Scientific Talent* (Princeton, NJ: Princeton University Press, 2014); Yi Xue and Richard C. Larson, "STEM Crisis or STEM Surplus? Yes and Yes," *Monthly Labor Review,* U.S. Bureau of Labor Statistics, May 2015, https://doi.org/10.21916/mlr .2015.14. See also Noah Weeth Feinstein, Sue Allen, and Edgar Jenkins, "Outside the Pipeline: Reimagining Science Education for Nonscientists," *Science* 340, no. 6130 (2013): 314–317. David F. Labaree offers insightful discussions of the conflicting goals of public education in *Someone Has to Fail: The Zero-Sum Game of Public Schooling* (Cambridge, MA: Harvard University Press, 2010).

9. Thomas F. Gieryn makes this point for science as a whole in *Cultural Boundaries of Science: Credibility on the Line* (Chicago: University of Chicago Press, 1999), 1–25. A more detailed version of this argument is made in John L. Rudolph, "Portraying Epistemology: School Science in Historical Context," *Science Education* 87 (2003): 64–79.

10. Steven Shapin, "Science and the Modern World," in *The Handbook of Science and Technology Studies,* 3rd ed., ed. Edward J. Hackett, Olga Amsterdamska, Michael Lynch, and Judy Wajcman (Cambridge, MA: MIT Press, 2007), 433–448; National Research Council, *Using Science as Evidence in Public Policy* (Washington, DC: National Academies Press, 2012); Chris Mooney, *The Republican War on Science* (New York: Basic Books, 2006); Thomas M. Nichols, *The Death of Expertise: The Campaign against Established Knowledge and Why It Matters* (New York: Oxford University Press, 2017).

11. The disproportionate emphasis on the process of sense-making over understanding the scientific enterprise in NGSS can be seen in Christina V. Schwarz, Cynthia Passmore, and Brian J. Reiser, eds., *Helping Students Make Sense of the World Using Next Generation Science and Engineering Practices* (Arlington, VA: NSTA Press, 2017). On expertise in society, see Harry Collins and Robert Evans, *Rethinking Expertise* (Chicago: University of Chicago Press, 2007), and Harry Collins, *Are We All Scientific Experts Now?* (Cambridge: Polity Press, 2014).

12. John L. Rudolph, "Inquiry, Instrumentalism, and the Public Understanding of Science," *Science Education* 89 (2005): 803–821. On useful truths, see Philip Kitcher, *Science, Truth, and Democracy* (New York: Oxford University Press, 2001).

13. For details of this approach, see John L. Rudolph and Shusaku Horibe, "What Do We Mean by Science Education for Civic Engagement?," *Journal of Research in Science Teaching* 53 (2016): 805–820. As similar critique is made in Per Kind and Jonathan Osborne, "Styles of Scientific Reasoning: A Cultural Rationale for Science Education?," *Science Education* 101 (2017): 8–31.

14. Joseph J. Schwab, "The Teaching of Science as Enquiry," in *The Teaching of Science,* ed. Joseph J. Schwab and Paul F. Brandwein (Cambridge, MA: Harvard University Press, 1962), 102, 71.

ACKNOWLEDGMENTS

Over the years since this project began as a loose set of ideas, I've had the opportunity to work with many individuals and institutions that have been crucial in helping me transform those ideas into this book. None of it could have been accomplished, of course, without the primary source documents, papers, and notes that give us a window into the past. I thank the archivists and librarians at the National Archives, National Academy of Sciences, Library of Congress, Caltech, University of Nebraska at Lincoln, University of Rochester, University of Michigan, Cornell University, Columbia University, Wisconsin Historical Society, and University of Chicago. I extend my appreciation especially to the archivists at MIT, Harvard University, and the Biological Sciences Curriculum Study in Colorado Springs, where I spent a considerable amount of time going through materials. Amy Crumpton and Norma Rosado-Blake at the American Association for the Advancement of Science deserve special mention for their help and hospitality during my trips to examine the papers of Project 2061 in Washington, DC.

At various points along the way I benefited from a National Academy of Education / Spencer Foundation Postdoctoral Fellowship, a Spencer Foundation Small Grant, the Graduate School at the University of Wisconsin-Madison, and the National Science Foundation (grant no. SES-0114542). This financial support has been greatly appreciated. Any opinions, findings, and conclusions or recommendations expressed in this material, however, are my own and do not necessarily reflect the views of the National Science Foundation or any of the other funding agencies.

I have had the advantage over the years of working with fantastic colleagues who, if not directly involved in shaping the content of this book, have nonetheless contributed to shaping me as a scholar and writer. Thanks to historians Jonathan

Zimmerman (who has always been a gracious supporter of my work), Ron Numbers, Lynn Nyhart, Dave Kaiser, Kathy Olesko, Chris Ogren, and sociologist Daniel Kleinman. Among the many science education scholars deserving mention are Mark Windschitl, Mike Ford, Bill Sandoval, Greg Kelly, Sherry Southerland, John Settlage, and Doug Larkin. Closer to home, I'm grateful to have worked with Peter Hewson, Melissa Braaten, Noah Feinstein, Leema Berland, and Rosemary Russ. They have continued to challenge my thinking about science education as well as kept things running smoothly in the Department of Curriculum and Instruction at Wisconsin while I was at work on this book and tending to my duties as department chair.

Thanks to Rick Duschl for feedback he provided on key chapters as well as for all he's done to encourage and support my work over the years, and to the others who provided comments on the draft manuscript, professional advice, and friendship along the way, in particular Sevan Terzian, David Meshoulam, Rich Halverson, Leema Berland, Audra Wolfe, Angelo Collins, and the anonymous reviewers for Harvard University Press. All errors or faults remaining in the book are mine, of course. Additional help was provided along the way by graduate students Drew Joseph, who provided feedback on the conclusion, and Eric Luckey, who served as an able research assistant and endnote expert. Dana Freiburger was invaluable in securing the images and permissions to use them in the book. His expertise in this area is unparalleled. Deirdre Mullane did a wonderful job helping me craft the book proposal to send to publishers as well as prompting me to think about how to reach a broader audience, and Jeff Dean, my editor at the Press, was tremendously supportive and encouraging throughout the process.

Thanks also to Mark Chandler for his willingness to discuss issues related to science, science education, and middle age, and to the rest of the card club in Madison, Wisconsin, for providing a much-needed outlet for relaxation and entertainment over the years. Greg Wold, Carla Dillman, Don Drott, and Judy Christensen deserve mention for their continued support and friendship.

Special thanks go to Adam Nelson, who offered invaluable feedback on the manuscript and has provided support, inspiration, and comradery over the years, and Bill Reese, who has been a loyal colleague and friend. Our conversations about scholarship, writing, and life over a pint or two have been more helpful than he can ever know. Jim Stewart helped get all this started in so many ways, and I remain ever grateful for our friendship. Finally, none of this would have been accomplished without the love and support of my family—my mother, who unfortunately passed away before this project was completed, my wonderful daughters Audrey and Lydia, and Jen, my wife, who has given me everything I could ever ask for and more than I deserve. Thanks to Stella as well, who recently joined our family and was with me every day during my sabbatical as I completed the manuscript.

INDEX

Committee on General Education in a Free Society, 118, 130
Committee on Secondary School Studies. *See* Committee of Ten
communism, 138, 139, 147
Compton, Karl, 124, 145
Conant, James Bryant, 131–137, 141, 142, 145, 155, 176, 193, 195, 230; advocacy of the history of science, 132–134, 137; and the Harvard Redbook, 118; and the tactics and strategies of science, 132, 156; Bampton Lectures, 135, 145; conferences on science in general education, 134–135, 156, 187; critique of scientific method, 130–132, 144; general education program in science at Harvard, 131–134, 137; and Natural Sciences 4, 133–134; summer course for high school teachers, 135; Terry Lectures, 131
Cook County Normal School, 50
Cooke, Josiah, 27, 37, 38
Cooley, Edwin, 91–92
corn clubs, 112
Coulter, John G., 93
Coulter, John M., 88, 110
country life movement, 112
creation science, 199
creationism, 199–201
Culture Demanded by Modern Life, 19

D. C. Heath and Company, 15, 91
Darwin, Charles, 17, 28–29, 46
deductive reasoning, 28–29
Dewey, John, 11, 72, 103–104, 122, 135–136, 192, 193, 225; AAAS vice-presidential address, 80–81; and general science, 109–110; and the laboratory school at Chicago, 87–88; and reflective thought, 89–91; as leader of educational reform, 85; association with step-by-step scientific method, 143; complete act of thought, 90, 92, 95; efforts to correct public misunder-

standing of scientific method, 136; influence on Joseph Schwab, 151; influence on science education, 88, 94–96; instrumental view of science, 86, 229; view of scientific method, 81–82, 90–91; views on science, 88; work with scientists, 88, 90
Dodge, Charles W., 48
Dollard, Charles, 135, 136
DuBridge, Lee, 145
Dull, Charles, 101
Dunham, Alden, 189, 223

Earlham College, 142
Eastern Association of Physics Teachers, 59–62
Eastman, Max, 95
economics, 196
education, professionalization of, 76
educational psychology, 58–59, 63–65, 71, 75
Edwards v. Aguillard, 199
Eikenberry, William, 105–111
Eisenhower, Dwight, 185
Elements of General Science, 107, 108
Eliot, Charles, 11, 29, 30, 31, 44, 56, 64; advocacy of science in the college curriculum, 20; and laboratory teaching, 37; and the elective system, 20; as chairman of the Committee of Ten, 49–52, 58–59
environmentalism, 182
environmental science, 114
evolution, 17, 28, 46, 65, 185; and creationism, 199; teaching of, 199, 201

faculty psychology, 30–31, 71, 75
Faraday, Michael, 19
Forbes, Stephen, 26, 33; view of scientific method, 82–84
Fourteen Weeks series, 23–25, 36, 44
Friedman, Thomas, 203, 218
functional psychology, 85, 86, 89